# WHAT THE
# TUDORS & STUARTS
# DID FOR US

# WHAT THE
# TUDORS &
# STUARTS
## DID FOR US

## ADAM HART-DAVIS

B⬚XTREE

First published 2002 by Boxtree
an imprint of Pan Macmillan Ltd
Pan Macmillan, 20 New Wharf Road, London N1 9RR
Basingstoke and Oxford
Associated companies throughout the world
www.panmacmillan.com

ISBN 07522 15086

9 8 7 6 5 4 3 2 1

A CIP catalogue record for this book is available from
the British Library.

Inside design by Dan Newman, Perfect Bound Ltd
Colour reproduction by Aylesbury Studios
Printed and bound by Bath Press

# To Sue, with love

# By the same author

*Don't Just Sit There!* 1980
*Where There's Life...* (with Hilary Lawson), 1982
*Scientific Eye*, 1986
*Mathematical Eye*, 1989
*World's Weirdest 'True' Ghost Stories*, 1991
*Test Your Psychic Powers* (with Susan Blackmore), 1995
*Thunder, Flush and Thomas Crapper*, 1997
*Science Tricks*, 1997
*The Local Heroes Book of British Ingenuity* (with Paul Bader), 1997
*Amazing Math Puzzles*, 1998
*More Local Heroes* (with Paul Bader), 2000
*Chain Reactions*, 2000
*What the Victorians Did For Us*, 2001
*Henry Winstanley and the Eddystone Lighthouse* (with Emily Troscianko), 2002

# Acknowledgements

I am most grateful to the many people who have helped in the preparation of this book, especially the BBC team Alom Shaha, Billie Pink, Caroline van den Brul, Dan Kendall, Emma Sutton, Nikita Lalwani, Page Shepherd, Paul Bradshaw, Paul King, Steve Wilkinson, and Tanya Batchelor. Dr Allan Chapman provided many of the stories in the first place, and found time to read part of the text and correct several mistakes. Professor David Daniell provided helpful suggestions about William Tyndale. Mandy Little provided welcome encouragement throughout. My editors Emma Marriott, Siobhan Gooley and Christine King greatly improved and smoothed my rough text. Dan Newman did a wonderful job with the design, and made the book look beautiful.
Thank you all.

Being the
# CONTENTS
of Doctor Adam Hart-Davis's most eſtimable book,
## WHAT THE TUDORS & STUARTS DID FOR US

Publiſhed in LONDON by Boxtree, 2002.

# Introduction

y previous BBC television series looked at what the Romans and Victorians did for us. The Romans were fantastic at fighting, engineering and administration, and the Victorians likewise were brilliant engineers, with the advantages of steel and steam engines. For this later series, I knew the Stuart period was notable for scientific advances, but what could the Tudors have done for us? When they took over England in 1485, we were still in the Middle Ages, having lost most of those Roman skills a thousand years before. What's more, old ideas had trickled through from ancient times, and had often been wrapped in superstition and mystery, while new ideas were generally stamped on as heretical. Nothing like 'science' existed.

And yet, against the odds, new ideas did emerge, and by the time of the last Tudor, Elizabeth I, the foundations were being laid for the development of modern science. To begin with, many of these ideas came from the continent of Europe – the Dutch and the Italians seem to have been particularly ingenious – but they were rapidly taken up in Britain.

This book explores the subjects covered in the BBC television series, and it goes further in some directions, including material from diet and lifestyle through to clothes and personal hygiene. Most history books of this period concentrate on the politics, which were intricate and fascinating, but not my target. Instead I have looked for the science and the technology, and politics creep into this book only where there is a direct connection with the technology or the science. Thus Henry VIII was so desperate for a son and heir that he divorced Catherine of Aragon and therefore cut himself off from the Catholics and Rome. This had immense and far-reaching consequences, many of which are still with us today. Immediately he had to build up his defences against a possible invasion by France or Spain. As a result he designed new castles, developed new cannon technology, and set up the first standing navy. The most significant result of this military activity was the rout of the Spanish Armada in the reign of Henry's daughter Elizabeth.

As well as the material developments of the time, a new type of thinking was born, challenging existing orthodoxy. Religion was thrown into turmoil, partly by the spread of the Protestant movement but also by the translation of the Bible into English. For the first time, people who did not understand Latin were able to read the holy texts for themselves. In fact the development of the printing press both reflected and accelerated revolutions in understanding. The human body was properly examined and explained for the first time, and the heavenly bodies were seen in an entirely new light as astronomers made better observations and applied mathematical tools to their results.

The Tudors and their European contemporaries invented not only the pencil but also the first instruments to help with accurate drawing, which brought forward a whole new realism in art, and a more scientific approach to maps. This was typical of what seemed like a small idea leading to a great change in transmission of information and understanding. Better maps and better instruments led to more accurate navigation, and great sea voyages. The Tudors began to see the rest of the world, and a new place called America appeared on their maps. Indeed they settled a colony there, and brought back the first detailed news of exotic people and animals.

Meanwhile at home the more prosperous classes enjoyed themselves with fine houses, fine clothes, fine food, and entertainment of all sorts from music to theatre. Tudors invented the knitting machine, the shopping mall, and even the first self-contained water-closet.

In fact Henry's split from Rome seemed to provide the catalyst for all sorts of free and revolutionary thinking; the result was that by the end of the Tudor dynasty, England was already beginning to overtake the rest of Europe in scientific understanding. The great Italian scientist Galileo was one of the most outspoken pioneers of radical thinking about the natural world – he started making waves in the 1580s, and hardly stopped before his death in 1642 – but the philosopher Francis Bacon formalized much of the new thinking in the early 1600s, and his ideas were picked up throughout the seventeenth century; they led ultimately to modern science.

The first Stuart king was James I, who was already James VI of Scotland, and his first task was to try to unite the kingdoms and organize the island. Though not in the same league as the Romans, the Stuarts were great organizers, and produced the first union flag, the first public transport, the first royal mail, the first road maps, the first agricultural machinery, and even the Bank of England, which was set up to lend the government money – a loan that was never repaid.

Within a more organized society, the way was open for Bacon's scientific ideas to take root and flourish, and experimental science began to take off. William Harvey worked out how the blood circulated in the body, and, once Galileo had seen the moons of Jupiter, a galaxy of astronomers sorted out the circulation of the bodies in the sky. Edmond Halley in particular, after a tedious trip to the southern hemisphere, observed 'his' comet and made spectacular deductions and predictions. He also invented an improved diving bell, which along with the first submarine made underwater exploration far more plausible.

For ordinary people life was 'poor, nasty, brutish and short', and after the horrors of the civil war in the 1640s came the Great Plague and the Great Fire of London. Nevertheless from the ashes of the Fire grew the new London, built by Christopher Wren, the amateur architect who had learned from the great Inigo Jones. Actresses appeared on the stage, politics and newspapers took off in the newfangled coffee-houses, alchemy began to turn into chemistry on the cliffs of north Yorkshire, and a stubborn eccentric built an extraordinary lighthouse on the Eddystone Reef.

Finally all those burgeoning scientific ideas began to flow together. Isaac Newton sorted out the colours of the rainbow, and is also credited with discovering the laws of gravity, although recent evidence suggests that he made up the story of the apple many years after it was supposed to have happened. Meanwhile a chain of people, worrying about air, and air pressure, and steam, invented the air pump, the barometer, the pressure cooker and finally the steam engine. This engine was was the first machine capable of providing portable power, and so marked the beginning of the Industrial Revolution: this really was the appliance of science.

Filming the series was great fun. I had to dress up quite a lot, in rather tight and very hot clothes – I don't know how those Tudors survived the summer. On one particularly enjoyable day we chartered the Matthew, a replica of John Cabot's ship, and I discovered America. Well, at least they told me it was a new-found land…

As usual during the filming, I found myself on a steep learning curve; every day I was presented with new stories and new ideas, and knitting them into some sort of narrative has been an enjoyable challenge. One day I literally found myself knitting, as I did my best to understand the mechanism of the world's first knitting machine. This has been a fascinating task for me; I hope you enjoy the result.

Adam Hart-Davis
August 2002

# What the Tu

ors did for us

# The Tudor dynasty

The founder of the Tudor dynasty in England was Henry VII. Of Welsh stock, his claims to the throne were pretty shaky: his mother's great-great-grandfather had been

Edward III, and his father had been a half-brother of Henry VI. But he won a decisive victory over Richard III at the Battle of Bosworth in 1485 – the story goes that after Richard's death, the crown was found under a gorse bush and placed on Henry's head. The new king wasted no time in marrying the beautiful Elizabeth of York, uniting the houses of York and Lancaster and so ending the Wars of the Roses, which had been rumbling on for thirty years. A period of relative peace and prosperity began.

Henry and Elizabeth had two sons: Arthur, who died suddenly at the age of fifteen, and Henry, who succeeded to the throne when his father died in 1509. Henry VIII immediately married his brother's widow, Catherine of Aragon, who in twenty years bore him only one surviving child, Mary. Desperate for a son, he divorced her – bringing about the cataclysmic split from the Catholic Church – and married Anne Boleyn, who bore him another daughter, Elizabeth. Three years later he accused Anne of having other lovers, including her own brother, and had her beheaded on 19 May 1536. Within two weeks Henry married one of her maids, Jane Seymour, who produced his only surviving son Edward, but she died twelve days after the birth. He then hunted around for a fourth wife, married Anne of Cleves, but disliked her on sight and divorced her after seven months. Then Henry married Anne's pretty maid Catherine Howard, had her beheaded for sleeping around, and finally married Catherine Parr, Lady Latimer, who seems to have brought the whole family a measure of peace.

Henry died in January 1547; his sickly son Edward VI – the first Protestant monarch – was just nine and survived only until the age of fifteen. The Protector of Northumberland rushed to put his own daughter-in-law, Lady Jane Grey, on the throne, but she was deposed after just nine days in favour of Henry's daughter, 'Bloody' Mary – an ardent Catholic who wanted to bring the country back to the old faith: she had three hundred Protestants burned at the stake.

Mary reigned from 1553 until 1558; despite her desperate longing for a child, she was unable to conceive, and when she died her half-sister, the Protestant Elizabeth, took over. Elizabeth had survived a precarious early life, when she was often seen as a threat to the Catholic succession, and remained queen for forty-five years. Her powerful navy dominated the western world, through an age of adventure, discovery, the flowering of literature, and the first hints of modern science.

Despite the frantic efforts of Henry VIII to secure the Tudor succession, none of his three legitimate children had an heir; so when Elizabeth died the throne passed to her Stuart cousin, James. The son of Mary Queen of Scots, whom Elizabeth had had executed, he was already King James VI of Scotland; he now became King James I of England.

# A selective family tree

HENRY VII   m.  Elizabeth of York
(r. 1485–1509)

Arthur
(d. 1502)
m. Catherine
of Aragon

HENRY VIII
(r. 1509–47)

Margaret        Mary
(d. 1541)       (d. 1533)
m. James IV
of Scotland

m.(1)          m.(2)        m.(3)       m.(4)        m.(5)         m.(6)          James V
Catherine      Anne         Jane        Anne of      Catherine     Catherine      of Scotland
of Aragon      Boleyn       Seymour     Cleves       Howard        Parr           m. Marie of
(d. 1536)      (ex. 1536)   (d. 1537)   (d. 1557)    (ex. 1542)    d. 1548        Lorraine-
                                                                                  Guise

MARY I
(r. 1553–58)
m. King
Philip of Spain

EDWARD VI
(r. 1547–53)

Mary            Lady
Queen           Jane
of Scots        Grey
(ex. 1587)      (ex. 1554)

ELIZABETH I
(r. 1558–1603)

JAMES (STUART)
VI of Scotland
I of England
(r. 1603–25)

Being the First Chapter

# THE THINKING REVOLUTION

'I have practysed and lerned at my
grete charge & dispense to ordeyne
this said book in prynte... that every
man may have them att ones.'
WILLIAM CAXTON

*Previous spread: The title
page of the Great Bible of
1539 shows Henry VIII on
a vast throne, generously
handing out the word of
God (verbum dei).*

Åfter the turbulence of the Wars of the Roses, England entered a period of relative calm and confidence. When he came to the throne in 1485, Henry VII turned out to be an astute and careful administrator; he began to put his new realm in order, and people could turn their attention to other matters than surviving a civil war. They could take more notice of new ideas that were beginning to trickle in from the continent of Europe. As the fifteenth century drew to a close, radical thinkers were boldly challenging long-held orthodoxies: the Dark Ages were giving way to the Renaissance, when the world was 'reborn', or at least rediscovered. The legacy of the sixteenth-century revolution in thinking – in the sciences, in religion, and in technology – is with us still.

The Tudors were to see old certainties being shattered. People had always believed that the Sun, the Moon and the fixed stars all revolved around the Earth, and yet suddenly an astronomer suggested the unthinkable: that the Earth was not really the centre of the universe. The ancients had said that the heavens were forever unchanging – and yet suddenly a new bright star was seen. Everyone knew that the King and the Church and the Pope were supreme rulers, and yet suddenly the King wanted divorce, and took the Church in England away from the Pope.

The infernal machine that helped to bring about all these shifts was the printing press. For thousands of years, books had been written out by hand, copied painfully and laboriously by scribes. The great library at Alexandria, around 200 BC, had been built up by borrowing all the known great works, from Athens and elsewhere, getting them copied, and then disgracefully returning the copies and keeping the originals. Copying books was so slow and expensive that it could be done in only a few places; most of the scarce books in the world were to be found in monasteries and a few other libraries. The printing press would change all that.

# A revolution in communication

Primitive printing, with fixed woodcuts, began in China before AD 200; the Chinese went on to develop movable type by the eleventh century, but

this method did not reach the West until the 1440s. Then, the German Johannes Gutenberg reinvented printing with movable type, using metal rather than wood. He made for himself a stack of pieces of metal, or 'types', each with a single letter of the alphabet or other character, which he could then put in any order into a frame that fitted into his press. Thus he could print any word or sentence or book that he wanted. And he could make many copies of the same book, copies that would come

out identical, without all the mistakes that had been inevitable when books were copied by hand. One of the first books he produced, in the early 1450s, was a Bible – the Gutenberg Bible – of which a few copies still exist. This in itself is a tribute to the power of his invention: that one of its first products, made before any of the countless refinements that followed, should still be with us 550 years later.

Gutenberg was a goldsmith too, and he may have dabbled in alchemy (of which more later), for to get his movable type to work, he needed a clever stroke of what we should now call chemistry – but chemistry as a formal, separate discipline did not exist then. He latched on to the key metal: antimony. Antimony had been known for hundreds of years, and was used

*There isn't a Tudor printing press in existence; this replica was made for the BBC series.*

*A page from the 1455 Gutenberg Bible, one of the first books printed with movable metal type.*

in the fifteenth century as tartar emetic – it made you vomit instantly, which was then seen as a way of expelling bad humours from the body. Antimony metal pills were also used as reusable laxatives: swallow one and it would cause enough irritation in the gut to produce violent action, whereupon you could retrieve the pill for future use. There are even tales of these pills being used in families for generations. Like many such 'remedies', antimony is poisonous; several Victorian doctors murdered wives and relatives with antimony, and Mozart may have died of accidental antimony poisoning.

What mattered to Gutenberg, however, was the fact that antimony is the only metal that expands when it solidifies. All the others shrink. That means that if you pour molten antimony into a mould and let it set, the metal forces its way into every corner of the mould, and you get a superbly sharp casting. Thus the letters he made using antimony type were much sharper and clearer than could be made using just lead. In order to set a line of type, he had to fit together all the different letters to go on the line, and then put them together into a metal frame, traditionally called a galley. All the types had to be exactly the same size and shape to fit in the galley, and exactly the same height if all the letters were to put ink on to the paper. Gutenberg discovered that a mixture or alloy of 60 per cent lead, 10 per cent tin and 30 per cent antimony was ideal. It formed lovely sharp castings, and

*A lump of antimony, the metal that was vital to the technology of printing with movable metal type, and also made into reusable laxative pills.*

'The priests of the country be unlearned, as God knoweth there are a full ignorant sort which have seen no more Latin than that they read in their portesses and missals which yet many of them can scarcely read.'
WILLIAM TYNDALE

neither expanded nor contracted on solidifying, so that all the types came out the same size and fitted neatly into the galley.

The Gutenberg Bible was printed in Mainz, but the idea of printing with movable type quickly spread, and within a dozen years there were presses all over Europe and beyond. The technique was eventually brought to England by William Caxton, who set up a printing press in the precincts of Westminster Abbey around 1476. Caxton was a successful businessman, diplomat and man of letters. The first thing he printed was his own translation from the French of a book called *Recuyell des Histoires de Troy*, and he said he had to print it because the demand for copies was so great and the labour of copying them by hand so daunting: 'I have practysed and lerned at my grete charge & dispense to ordeyne this said book in prynte… that every man may have them att ones.' He printed his translation in Bruges around 1474, and then produced *The Game and Playe of the Chesse*, apparently the second book ever printed in English, and the first to use many woodcuts.

Caxton lived until about 1492, and produced around a hundred books – a tremendous output, and a great variety of material, ranging from encyclopedias and classics to romance and poetry. He worked for Edward IV, Richard III and Henry VII; so he had plenty of royal patronage, without which he could not have kept printing, since there were not enough other wealthy customers for his erudite books. He was not a revolutionary, but the printing press caused a revolution in communication.

For the first time, anyone could write a book, on any subject, and get it mass-produced. For the first time, anyone could read a book, without what amounted to censorship by the Church. As we shall see, the established Church was quick to condemn this democracy of communication – and, as far as it was concerned, with good reason: heretical ideas were being circulated. Few people were able to read, but the fact of printing allowed ideas to spread widely. In recent times, printing has been augmented by other means of communication: the telephone, the photocopier, the fax machine, the radio and television – and the internet is causing a further revolution in communication. Nevertheless, even the internet may not in retrospect bring such a change to society as was introduced by the printing press.

When Caxton died, his business was taken over by his assistant, Wynkyn de Worde, who wanted to make money from it, and therefore concentrated on cheap books for the mass market instead of expensive books for wealthy patrons. In 1500, in order to be nearer the booksellers, he moved the business to Fleet Street, from where he produced little volumes of popular romances – *Bevis of Hampton* and *Sir Degare* – potted histories, sensational news, DIY books, books of instruction for pilgrims, on manners, and on household practice, and books for children. He also printed school books, all sorts of religious books, and in 1523 the Lord's Prayer in English – a daring act at the time, given the Church's suspicions.

## Translating the word of God

Wynkyn de Worde, with his press in Fleet Street, could be called the founder of the popular printing industry in England, but a man who had far more influence on the English language was William Tyndale. Born in Gloucestershire in the 1490s, he studied at both Oxford and Cambridge, and became a skilled linguist. He went back to the West Country and began preaching, but met a hostile reception. Because he had studied the Bible in its original languages, he knew a great deal more about it – and what it meant – than the country vicars, who scarcely understood the version they had, which was in Latin, and a thousand years old. He later wrote: 'The priests of the country be unlearned, as God knoweth there are a full ignorant sort which have seen no more Latin than that they read in their portesses and missals which yet many of them can scarcely read... when they come together to the ale house, which is their preaching place, they affirm that my sayings are heresy.' This opposition in the established Church was to grow, and lead Tyndale to an awful fate.

*William Tyndale, whose brilliant translation of the Bible into English from the original Greek and Hebrew led to his execution for heresy.*

He decided that he would make it his life's work to translate the Bible from the original straight into English, so that anyone could read it. He said to a learned man: 'If God spare my life, ere many years I will cause a boy that driveth the plough shall know more of the scripture than thou dost.' Tyndale went to London, and tried to get support for his great project from the Bishop, Cuthbert Tunstall, but although Tyndale stayed there for a year, Tunstall refused to help. So Tyndale went off to Germany, staying in Hamburg and in Antwerp; but he began printing his New Testament in Cologne in 1525, after translating it directly from the original Greek. When a local Catholic took out an injunction to stop the presses, he moved to Worms, and finished the work there.

Copies of his New Testament were smuggled into England, hidden in bales of cloth, and as soon as they began to circulate were condemned by the Church, and banned by royal proclamation in 1530. Henry VIII, who had come to the throne in 1509, was still an enthusiastic Catholic, endorsing the authority of the Church, and he disapproved of dangerous and inflammatory material. Powerful churchmen did not want ordinary people to get their hands on the gospels, because this would seriously undermine the authority of the priests, the Catholic hierarchy. Readers might find contradictions and absurdities in the holy texts, which would cause further problems for the

Church. For example, they might read in one of St Paul's epistles: 'For by grace are ye made safe thorow fayth and that nat of yourselves. For it is the gyfte of God and commeth nat of workes...' In other words, salvation depended on the depth of your faith, rather than the amount of money you gave to the Church! Not a popular view among churchmen.

So the authorities regarded Tyndale's aim to share the received gift of God with everyone as a terrible heresy. Even Tunstall, once Tyndale's friend, called the book 'pestiferous and pernicious poison' – and he burned it. Tyndale was shocked. A bishop had burned not merely his book, but the words of God, lovingly translated by a scholar from the original Greek.

Tyndale carried on with his great plan, and began translating the Old Testament from the original Hebrew; the first book, Genesis, was printed in 1530. He had completed about half of the Old Testament when he was betrayed, arrested and tried. He was executed in October 1536 at Vilvorde, just north of Brussels; he was strangled and then his body was burned. This was the Catholic Church's response to a man whose 'crime' had been to bring the word of God to the people of Britain. Tyndale's last words were: 'Lord, open the King of England's eyes!'

# A revolution in religion

When Tyndale was in Wittenberg, he had met Martin Luther, an outspoken priest who once claimed he could drive away the evil spirit with a single fart. Luther had visited Rome, and had been sickened by the luxury and the political machinations at the Vatican; he had developed a violent dislike of some forms of popery, and in particular the sale of indulgences (with which sinners could be forgiven their sins for a fee). He preached a simple doctrine based on the teachings of St Paul and St Augustine, and for him the sale of indulgences was the ultimate abuse. In 1517 he nailed to the church door a list of forty-nine reasons why the sale of indulgences was appalling, and this protest began the Protestant movement. Its spread was helped enormously by both the printing press itself, and the publication of biblical texts in translation – for the heart of Protestantism was the Bible as the sole source of God's truth.

## *The king's 'great matter'*

At first, Henry VIII was strongly opposed to Luther's teaching; in 1521 he wrote – with some help from his friends – a vigorous attack on him, called *Assertio Septem Sacramentorum*. This book became a bestseller all over Europe, and the Pope was so pleased he created a new title for Henry: Fidei Defensor, or Defender of the Faith. By English Act of Parliament this title was declared hereditary; all subsequent monarchs have used it, and even today British coins are stamped F.D.

However, Henry's devotion to the Church was to be tempered by his desperate desire to produce a son and heir, who would ensure the Tudor succession. His wife, Catherine of Aragon, was loving and faithful but, after a series of miscarriages and stillbirths, she had only one surviving child: a girl, Mary, born in 1516. As the years rolled by, Henry became more and more certain that Catherine would not produce a son, and by 1530, when she was forty-four years old, he gave up hope. Frustratingly, his mistress

*As a young man, Henry VIII was tall, athletic and handsome.*

'I have changed my mind about having a barber to cut my hair; I think perhaps I shall allow the king to cut off my head and my hair at the same time.'

THOMAS MORE

Elizabeth Blount had given birth to a healthy boy in 1519 – the first of Henry's sons to survive, but undeniably illegitimate.

The crunch came in December 1532 as his girlfriend, Anne Boleyn, was pregnant. Could this be the son and heir he longed for? Divorce was not permitted by the Catholic Church, but Henry hit on an ingenious idea. He had married his brother's widow; this must be a form of incest and therefore not a real marriage. He had been living in sin all these years, and therefore was free to marry Anne. But the Vatican was hardly likely to agree. Even if he had approved in principle, the Pope was now under the influence of King Charles V of Spain. He was the nephew of Catherine of Aragon, and Charles was not going to allow his aunt to be casually discarded.

Henry now found he needed Tyndale, or at least his ideas. In 1528 Tyndale had published *On the Obedience of a Christian Man and how Christian Rulers Ought to Govern*. This too was swiftly banned by the Church, but Anne put a copy into his hands. In this book, Tyndale said that authority should not be divided between Church and State, and that a Christian king should be head of both State and Church. Here was a possible solution to Henry's problem: the Pope had refused to declare his marriage to Catherine of Aragon null and void, but Henry could simply declare himself Supreme Head of the Church of England, in effect a local Pope. This is exactly what he did. He persuaded Archbishop Thomas Cranmer to declare his marriage null and void, and on 25 January 1533 he married Anne Boleyn. However, she was already pregnant, and Cranmer did not declare his previous marriage null and void until three months later, in April.

Cranmer bent to the King's will, but the Lord Chancellor Sir Thomas More refused to do so. Even though he had been a close friend, Henry locked him up for fifteen months in a cold damp cell in the Tower because he would not agree that the king was above the law – and the law said that Henry was married to Catherine. Henry kept sending messengers to ask whether More had changed his mind, and on one occasion More said that he had. Lawyers were quickly summoned to take More's statement about the legality of the divorce, but More said, 'Oh no, I haven't changed my mind about that! I have changed my mind about having a barber to cut my hair; I think perhaps I shall allow the king to cut off my head and my hair at the same time.' And so it happened, on Tower Hill, on 6 July 1535, and More, still calm and joking, removed his beard from the block with the remark that 'it had never committed treason'.

Anne Boleyn bore Henry a daughter, Elizabeth, but no son, and even though he married four more times he managed to produce only one son, Edward, who died at the age of fifteen; so Henry's marital machinations, while causing the monumental split with the Catholic Church, never did ensure the succession.

## Henry's Great Bible

In 1538, six years after he instigated the split with Rome, and as part of his aim to define the Church of England, Henry decreed that there should be an English Bible in every church in the country. Tyndale's dying wish was granted: the King's eyes had indeed been opened, for suddenly the Bible in English was a Good Thing. With some revisions from Miles Coverdale, Tyndale's work became Henry's Great Bible, first printed in 1539, the first approved English translation.

On the title page of the Great Bible (see page 16), printed in two colours, is a fascinating piece of propaganda – the first printed attempt in England to put over a positive image of the monarch. At the very top (indeed, cut off in some editions) is the face of God among the clouds, and immediately underneath is Henry VIII himself, sitting proudly on a huge throne, and generously handing out bibles, while down below the great unwashed masses are shouting *Vivat Rex!* (Long live the King!). Ironically, each of the bibles being distributed is labelled *verbum dei* (the word of God): the same word of God that Tunstall burned and that had Tyndale executed for heresy. Even the renowned Authorized Version printed for James I in 1611 was mostly due to Tyndale. This became the best-selling book the world has ever seen.

Everyone in England went to church, and every week chunks of the Bible were read aloud. Where dialects had varied from town to town, gradually Tyndale's words and expressions became the standard form of English for the whole country. He displayed extraordinary skill in converting both Greek and Hebrew into clear, simple English – not for him the convoluted incomprehensible polysyllabic pronouncements typical of contemporary scholars, but crisp short words, easy to understand. He always used words of one syllable where he could: 'the poor and the maimed and the halt and the blind' comes from the Greek, while from the Hebrew he wrote 'And they heard the voice of the Lord God as he walked in the garden in the cool of the day...' and 'took a stone of the place, and put it under his head, and laid him down in the same place to sleep. And he dreamed and behold there stood a ladder upon the earth, and the top of it reached up to heaven...' This was language that anyone could understand. As he himself said, 'It was not possible to establish the lay people in any truth except the scriptures were so plainly laid before their eyes in their mother tongue.' Tyndale also gave us such phrases as 'the powers that be', 'let there be light', 'the beasts of the field' and 'the spirit is willing but the flesh is weak'.

We are said to speak the language of Donne, Milton and Shakespeare, but there is a good case to be made, supported especially by David Daniell's superb biography, that Tyndale's influence was greater even than that of

Shakespeare. Of all the things that the Tudors did for us, perhaps the most profound was to give us the English language, and the man who achieved that was William Tyndale.

# A revolution in mathematics

One man whose books had a considerable and lasting influence on people's thinking was Robert Recorde, who was born in Tenby, in south-west Wales, in 1510. His father was the mayor of what was then a busy port; the harbour was lined with big houses belonging to merchants. Robert studied at Oxford, where he became a fellow of All Souls College, and lectured on mathematics, rhetoric, music and anatomy. Later he went to Cambridge, and finally settled in London. He was apparently physician to Edward VI and to Mary, even though he was a Protestant and she a Catholic.

*A memorial to Robert Recorde in St Mary's Church, Tenby.*

### The equals sign

Recorde produced a string of books; his first, in 1543, was called *The Grounde of Arts*, and was a simple textbook of arithmetic, but so good it went through at least fifty editions. This was the first ever mathematics textbook in English, which may have had something to do with its success. In 1547 he produced a treatise on medicine, *The Urinal of Physicke*. He continued in 1551 with *The Pathway to Knowledge*, all about geometry, *The Castle of Knowledge*, in 1556, about astronomy, and finally in 1557, *The Whetstone of Witte*. In this last book he not only introduced plus and minus signs – the first time they had been used in English – but also invented the equals sign.

Apparently this sign used to be an abbreviation for *est* (Latin for 'is'), but he formalized it in his book, and said 'to avoid the tediouse repetition of these woordes: is equalle to: I will lette… a paire of paralleles, or Gemowe [twin] lines of one lengthe, thus: ====, bicause noe. 2. thynges can be moare equalle.'

Many of his books included dreadful verse – for example on the title page of *The Whetstone of Witte* he explains that we all need to keep our wits sharp:

*Though many stones doe beare great price,*
*The whetstone is for exersice,*
*As needefull, and in woorke so straunge,*
*Dulle thinges and harde it will so chaunge*
*And make them sharpe, to right good use…*

In Recorde's first maths book there are 2-, 3-, 4- and 5-times tables, but he reckoned the later ones were too difficult; so he taught a curious method of multiplication. If he wanted to multiply, say, 7 x 8, he first wrote one above the other:

>7
>8

Then he subtracted each from 10 and put the difference to the right of each digit:

>7 3
>8 2

He multiplied the two digits on the right to give the second digit of the answer:

>7 3
>8 2
>‾‾
>6

Then he subtracted diagonally (7-2 or 8-3) to get the first digit:

>7 3
>8 2
>‾‾‾
>5 6

It may seem a weird way of multiplying to modern eyes, but it actually seems to work.

He got into a serious argument with the Earl of Pembroke, and sued him, but unfortunately Pembroke was a close personal friend of Queen Mary and King Philip; so Recorde ended up in prison, and died there in 1558, just before Mary.

All Recorde's books were written in the form of dialogues between teacher and pupil, and sometimes they included entertaining problems. For example, the teacher asks the student: 'If a horse has four shoes, each with six nails, and you pay half a penny for the first nail, one penny for the second, two for the third, four for the fourth, and so on, doubling every time, how much will the horse cost?' The question is ambiguous – he could mean the shoes taken separately, or all together. If they are counted separately, the answer is 123 pence, which corresponds to 51p in today's money. If all the shoes are combined, then the answer is 8,388,607.5 pence, or £34,952.53 in today's money (which no doubt would have bought several stables of horses in the 1550s!).

Although he is remembered primarily as a mathematician and translator of mathematics, Recorde was interested in many other branches of learning, and one of the important things he did was to publish in English some of the ideas of Nicolaus Copernicus. More of these ideas were published by Thomas Digges, a mathematician, military engineer, and astronomer.

*Nicolaus Copernicus, who asserted that the Earth spins on its axis and revolves around the Sun.*

NON PAREM PAVLO GRATIĀ· REQVIRO
VENIAM PETRI NEQ. POSCO, SED QVAM
IN CRVCIS LIGNO DEDERAS LATRONI
SEDVLVS·ORO

# Heavenly revolution

Copernicus was a Polish astronomer who, just before he died in 1543, produced a book called *De Revolutionibus Orbium Coelestium* (*On the Revolutions of Heavenly Spheres*), in which he said that the Earth is not the centre of the universe, but revolves around the Sun, and the Earth is not even stationary in its orbit, but spins on its axis, which brings about day and night.

The Greek astronomer Aristarchus had suggested around 250 BC that the Earth revolves around the Sun, but unfortunately the Graeco-Egyptian astronomer and geographer Ptolemy, four hundred years later, ignored Aristarchus and produced his own theory, which came to be called the Ptolemaic System. Ptolemy placed the Earth at the centre of the universe, and said that everything else revolves around us.

At first sight, this seems obvious. Look up into the sky in the northern hemisphere, and you can see the Sun rising in the east, crossing the sky, and setting in the west. The Moon follows almost the same path, and all the stars and planets likewise rise in the east and set in

*A Tudor armillary sphere, showing the Earth at the centre of the universe, surrounded by rings representing the solid crystal spheres that carried the Moon, the Sun, each of the planets, and the fixed stars.*

the west. Clearly we sit still, and all the rest of the universe spins around us. There was a slight problem in that certain planets – Mars and Jupiter – occasionally seem to turn back for little diversions in their paths through the heavens – the Greek word 'planet' means 'wanderer' – but Ptolemy explained this by saying that they revolved in epicycles, like little rings spinning round the main hoops of their orbits. Although these epicycles were complicated and improbable, no one apparently realized that a much simpler explanation was possible.

In order to demonstrate the movements of the stars and planets, craftsmen built 'armillary spheres' in which the Earth was stationary at the centre, and all the heavenly bodies were fixed to intersecting metal circles like bracelets so that they could revolve independently around the Earth (*armilla* is Latin for bracelet). These armillary spheres were ingenious, beautiful, and utterly wrong.

Copernicus followed his predecessors in believing that the 'fixed stars' were all attached to a giant solid sphere outside everything else in the universe. He began his argument by pointing out that these 'fixed stars' are enormously far away from us, and therefore it is inherently unlikely

'The fool wants to reverse the whole science of astronomy, but sacred scripture tells us that Joshua commanded the Sun to stand still, not the Earth.'
MARTIN LUTHER ABOUT COPERNICUS

that the entire sphere of the fixed stars is spinning around every twenty-four hours. It is much more likely that the Earth is turning and the sphere is still.

He went on to prove mathematically that the Sun must be near the centre of the universe, and that the planets and the Earth revolve around the Sun. This 'heliocentric' system explains why Mars and Jupiter sometimes seem to go backwards: because the Earth moves faster than they do, we appear occasionally to be catching them up, so that, against the background of the fixed stars, they seem to be going backwards. It also explains why we always see the inner planets Mercury and Venus close to the sun. His assertion that the Earth is not the centre of the universe, but merely a planet, revolving around the Sun, was a powerful and probably heretical statement; it certainly contradicted the Ptolemaic view that had been accepted and blessed by the Church. For this reason, Copernicus made sure his book was not published until shortly before he died, and for good measure dedicated it to Pope Paul III.

Copernicus did not convince everyone. There was no way of testing his theory in the sixteenth century, and all the scholars could do was carry on arguing about whether or not his 'proof' was convincing. However, the publication of his ideas, especially in English, allowed this debate out into the open. Martin Luther said, 'The fool wants to reverse the whole science of astronomy, but sacred scripture tells us that Joshua commanded the Sun to stand still, not the Earth.' The Church might have preferred to suppress the Copernican theory, but the proliferation of printing presses was making censorship increasingly difficult.

## The man with the silver nose

One man who never completely accepted the heliocentric system was the Dane Tycho Brahe – surprisingly, for he was the most skilful astronomer of the sixteenth century. He was paid by King Frederick II of Denmark to observe the heavens from the observatory he built, Uranieborg (castle of the heavens), and later by the Holy Roman Emperor Rudolph II. Brahe was of noble birth, and in 1566 lost half his nose in a duel, after which he sported a false nose made of silver.

Brahe watched the courses of the planets far more precisely than anyone had done before. He observed a comet in 1577, and realized that it must have come from far beyond the Moon, so that it was apparently passing straight through the 'spheres' that held up the other planets; but these were supposed to be solid, and this was bad news for the Aristotelian and Ptolemaic theories. However, even more serious had been his 1572 observation of a new star in the constellation of Cassiopeia. He knew the sky so well that he was quite sure that it had not been there before. We now call what he saw a supernova, but in the sixteenth century this observation was rather scary. The ancients had said that the heavens were unchanging,

and had never changed since the beginning of time. This assertion had been accepted by the Church. How then could there possibly be a new star?

Brahe was making all his observations with the naked eye – the telescope as we know it was not invented until more than thirty years later – so a new star had to be extremely bright to be obvious even to a good observer. A supernova is a star that explodes and suddenly becomes millions of times brighter, which makes it far more easily visible. The most famous was the supernova observed by Chinese astronomers in AD 1054, whose remains are the Crab Nebula. A new one in 1572 was a nasty shock to the establishment.

*Tycho Brahe, the man with the silver nose, pictured here before the duel.*

## *Astrologer royal*

This supernova was also recorded by an extraordinary Englishman, John Dee, from his observatory in Mortlake, Surrey, working with his assistant Thomas Digges, who had translated the works of Copernicus. The new star was clearly visible through the winter of 1572, and the following year Dee wrote *Parallacticae Commentationis Praxosque*, in which he gave trigonometric methods that could be used to find out the distance to the supernova.

Born in 1527, Dee became a mathematician and an astrologer, and one of the first fellows of Trinity College, Cambridge, when the college was founded by Henry VIII in 1546. He spent much of his life wandering around Europe, and when he was in Britain fell in and out of favour with all the monarchs from Henry VIII to Elizabeth. Academically, he oscillated from serious mathematics and astronomy to alchemy and magic.

John Dee was given a pension by Edward VI, but lost his royal patronage when Edward died young in 1553. However, he remained in effect Astrologer Royal; Queen Mary summoned him to cast her horoscope. He immediately went and performed the same service for her sister Elizabeth, but Mary did not approve of this, and had him charged with treason and thrown into prison. When Mary died and Elizabeth became queen, the first thing she did was to ask Dee to work out which day would be most favourable for her coronation, and she accordingly held it on 15 January 1559.

Dee published books on alchemy, astrology and mathematics, but also one on explorers and exploring, entitled *The Perfecte Art of Navigation*. Indeed, he seems to have been adviser to some great sailors including Francis Drake, Martin Frobisher and Walter Ralegh. He is said to have contributed to the defeat of the Spanish Armada by putting a hex on the invaders, which produced the terrible weather they encountered in the North Sea.

For Dee, the occult and the world of the spirits were as real and as powerful as astronomy and the world of science. They were all part of the same struggle to understand natural phenomena, and if possible control them. He accordingly welcomed into his house during the 1580s a disreputable Irishman called Edward Kelly, who claimed he could commune with spirits and see the future in a crystal ball. Dee lived with this man and believed much of what he said, hoping that the spirits would tell him how to find the philosopher's stone that would turn things to gold, or to find buried treasure – until Kelly claimed the spirits had instructed them to swap wives! This was too much even for the genial Dee, and he sent Kelly packing.

Kelly may have been found lacking, but unorthodox views like Dee's were becoming increasingly tolerated in Britain. As the sixteenth century wore on, and the Protestants gained strength, so people opened their eyes to Renaissance ideas. One of the Renaissance men who was welcomed into England in 1583 was a curious, entertaining and fiery philosopher called Giordano Bruno. Originally an Italian, he spent most of his life wandering about Europe, denouncing Aristotle and preaching the ideas of Copernicus. He was also highly critical of established religions, laughing at miracles and putting the Jewish histories on a par with the Greek myths. Rashly, in 1593 he went to Venice, where he was put in jail, and in 1600 he was excommunicated and then burned at the stake for heresy.

Copernicus and Brahe were able to make improvements in astronomy, because of what had gone before: over the centuries and even through the Dark Ages (popularly thought to have begun with the collapse of the Roman empire in the fifth century AD), a number of people had studied the stars, and astronomy had gradually moved on. However, most of the other branches of what we now think of as science had stagnated since the time of the ancient Greeks. In many areas of study no one had even questioned what Aristotle had written around 330 BC.

# Anatomical revolution

Aristotle made beautiful observations, for example of the development of a chick inside an egg, but he did not do experiments. He said that the human body was composed of blood, fat, marrow, brain, flesh and bone,

*Meeting of medical minds: Galen and Hippocrates could never have met in real life, since Hippocrates, the 'father of medicine', died about 500 years before Galen was born.*

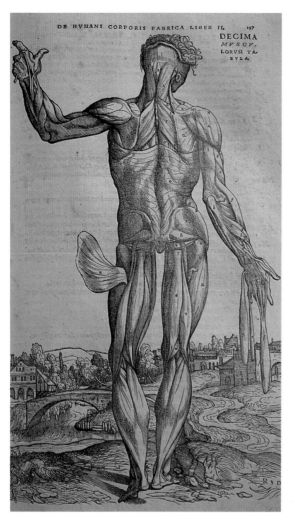

DE HVMANI CORPORIS FABRICA LIBER II. 197
DECIMA
MVSCV,
LORVM TA-
BVLA.

*Andreas Vesalius was the first person to publish details of the inside of the human body. His book* De Humani Corporis Fabrica *appeared in 1543.*

and that the main job of the brain was to cool the body by the production of phlegm. The science of anatomy took a big step forward 500 years later when the Greek physician Galen was summoned by the Roman Emperor, Marcus Aurelius, in the second century AD to be personal physician to his son in Rome. Galen also treated wounded gladiators. He did not dissect human corpses, but he did dissect all sorts of animals, especially goats, pigs and monkeys, to find out how various bits worked. He wrote about the functions of the kidney, the bladder, the arteries, and the valves of the heart.

Galen's theories were fundamentally wrong, but unfortunately he wrote rather well, and persuasively; when his texts were rediscovered in the early 1500s they were translated directly from the Greek into Latin, and everyone believed they were correct. The typical anatomy professor at Padua University was often a scholar of classics rather than medicine. He would sit on a high chair or *cathedra*, and read out the text (*ex cathedra*), while an inferior person, the barber-surgeon, would try to carry out the menial task of dissecting the corpse according to the instructions. Since the surgeons often did not understand Latin, the whole procedure could descend into farce.

The man who put anatomy back on a sensible course was Andreas Vesalius, who was born in Brussels but did his great work in Padua. He loved anatomy, and as a student frequently spent hours studying the bones in the Cemetery of the Innocents; he claimed that in half an hour he could identify any human bone while he was blindfolded. When he became a professor, he did all the dissections himself, which was so extraordinary that people came from far and wide just to see a professor getting his hands dirty.

In 1543, after four years of dissecting, Vesalius published his greatest book, *De Humani Corporis Fabrica* (*On the Structure of the Human Body*). His drawings showed for the first time, and in spectacular detail, the bones, the muscles, the nerves and the blood systems. To reproduce them, he hired the finest Venetian craftsmen to make woodcuts from pearwood soaked in linseed oil. Even today his pictures are spectacular, and back then they began a revolution in anatomy, just as Copernicus's book *De Revolutionibus Orbium Coelestium* – published in the same year – began a revolution in astronomy.

In order to complete his book, Vesalius had to dissect a large number of corpses, but luckily he was able to get help from a judge who let him have the bodies of executed criminals, and even occasionally delayed the execution to fit in with Vesalius's dissecting schedule. The book was so important that an English translation was printed only two years after the original, for the benefit of those who did not understand Latin.

Thirty years after Vesalius died in 1564 a beautiful anatomy theatre was built at Padua (see page 164), to allow many students to watch the professors in action. There was for a time a ban on dissecting human corpses, so look-outs were posted to watch for the authorities, and if necessary they quickly winched down the table with the human corpse and winched up in its place a table with a half-dissected pig, just to keep on the right side of the law.

*Dissection of a corpse was difficult for the barber-surgeon when the instructions were being read aloud in Latin.*

# Revolutionary medicine

However, human anatomy was far from the only subject in which fresh young radicals were beginning to question the wisdom of the ancients; the concepts of medicine were also changing. If you were ill during the Middle Ages you would probably visit an apothecary or a physician, who would first need to diagnose your problem. The body was thought to contain four 'humours': blood was the humour of the heart, phlegm the

*Doctors were often called piss-prophets because they examined a patient's urine to help diagnose disease.*

humour of the brain, yellow bile the humour of the liver, and black bile the humour of the spleen. A healthy body had these humours in balance, but if they got out of balance you would become ill.

So if you had a cold, then you clearly had too much phlegm, which was cold and wet; so the remedy would be something warm and dry – perhaps pepper and rocket seed. If you had a fever, on the other hand, you clearly needed a cool remedy, such as camphor. However, to check this diagnosis the physician would probably examine a specimen of your urine, and look for foam, cloudiness, curious bitter smell and unusual taste, which provided further clues about the balance of your humours. This is why medical men and apothecaries were often called 'piss-prophets'. John Collop wrote:

> Hence 'looking glasses', chamber pots we call,
> 'Cause in your pisse we can discover all.

A story goes that when Thomas More, Henry VIII's Chancellor, was languishing in the Tower because he would not agree that it was lawful for Henry to divorce Catherine of Aragon, he called for a 'looking glass', peed into it, and then held his urine up for inspection. 'Hm,' he said, 'I think I may live a few more years, if His Grace permits.' But Henry did not permit, and More was executed.

Whatever the diagnosis by the physician, the first part of the treatment was usually purging: you would be given a powerful emetic to make you vomit and a powerful laxative to clear out the rest of the tubes – a mixture of rhubarb root and agaric would probably do the trick. This was thought to get rid of all the excessive humours and leave you ready for the next stage of the treatment, which was often bleeding. You would be bled either with leeches or simply by having a vein cut until a cupful of blood had run out; this was reckoned always to be beneficial. Then they would get down to the more subtle business of trying to rebalance the humours.

Most of these treatments now seem odd, but some are downright bizarre; for example, an eye infection might be treated by raising blisters with crushed blister beetle and deadly nightshade, while for constant bed-wetting you might be given the dried windpipe of a cockerel.

The man who began to change this, and make treatment slightly more scientific, was the curious Philippus Aureolus Theophrastus Bombastus von Hohenheim, who was born near Zurich in 1493. He was inspired by the example of Aulus Cornelius Celsus, a Roman physician of the first century AD. Celsus wrote a medical encyclopedia based on the ideas of Hippocrates but tempered with practical experience. Reasoning that this was a step in the right direction, von Hohenheim rather pompously called himself Paracelsus – 'beyond Celsus'. He became Professor of Chemistry at Basel, and was a noisy and argumentative radical. He

'The physician's duty is to heal the sick, not to enrich the apothecaries.'
PARACELSUS

publicly burned the books of Galen, whom everyone still thought was the great genius. He announced to his students, 'You are not worthy that a dog shall lift his hind leg against you.' He called local doctors a 'misbegotten crew of approved asses', and the apothecaries' remedies 'foul brews', saying, 'The physician's duty is to heal the sick, not to enrich the apothecaries.'

All this made him thoroughly unpopular, and he was expelled from the university; however, he did plant the idea that disease might attack the body from the outside, rather than by imbalance of humours, and so the body might be helped to defend itself with minerals (that is, chemistry) rather than with herbal brews. He was the first to use tincture of opium in treatment; he tried compounds of iron, lead, zinc and arsenic, and he was the first to use mercury internally. Many of his treatments were poisonous, and no doubt they killed a number of patients, but they were often lethal to

*Philippus Aureolus Theophrastus Bombastus von Hohenheim, who called himself Paracelsus, was one of the founders of modern pharmacy.*

the disease too, and they certainly paved the way for a more scientific approach to medical treatment. In particular he rejected authoritarian doctrine, and encouraged original research.

# Alchemy to chemistry

Paracelsus was doing his best to use the new ideas of what we should call chemistry, which were developing from the ancient art of alchemy. *Chem*, meaning black, was a name for Egypt, and *chemeia* meant the strange skills the natives had there. In Arabic that became *al-chemeia*, or alchemy. The fundamental principles of alchemy were written down in Arabic around AD 900:

> *All materials are composed of a mixture of the four elements: earth, air, fire and water.*

> *The noblest of the metals is gold, followed by silver.*

> *Any metal can be changed (or transmuted) into any other metal by adjusting the proportion of the elements.*

> *This transmutation can be brought about by a particular agent.*

Some thought the fourth principle referred to what Aristotle called the fifth element; it was supposed to be a dry powder with magical healing powers. The Greeks called it *xerion*; in Arabic this became *al-ixir*, or elixir; later in English it became the philosopher's stone.

Through most of the history of alchemy its practitioners were trying to turn base metals, such as lead and mercury, into gold. However, there were other tricks that must have seemed like magic; for example when they heated calamine, a zinc ore, with copper they managed to extract a wonderful shiny yellow metal that we call brass, but it looked like gold, and they must have felt they were winning. They also managed to isolate a shiny silver metal called antimony, which turned out to have a strange, almost magical property that led to the possibility of printing with movable type (see page 19).

They were always trying to purify things, to break them down into their constituents, and one of their most important instruments was the still. They distilled everything from herbs to mercury, and sometimes achieved curious results. Thus if you break a couple of eggs, mix with a little wine, and distil the mixture, you can see the vapour condensing at the top of the flask, and you get first the 'aqua-vita' which is essentially a mixture of alcohol and water. This process was known as separating the 'spirit' from

the 'body' of the mixture – which is why brandy, whisky and suchlike are called spirits. Next you get some golden yellow 'oil of radish', and finally there distils over the least volatile part of the mixture, called 'castor oil', but the smell that accompanies this third stage is extremely unpleasant, because of the breakdown of the sulphur proteins in the egg.

*The pestle and mortar were essential tools in the curious and occult practices of alchemy.*

The complex mix of primitive chemistry, astrology, occult practices and downright magic that gave alchemy its potency gradually gave way to what would become modern chemistry, as the practitioners saw with clearer light. But certainly throughout Tudor times, alchemy was a valued basis for examining a mysterious world.

# Educational revolution

Meanwhile, education in England was undergoing a major upheaval, because after Henry VIII had dissolved the monasteries in 1535 and 1536, there was no one to run the schools. The same was true of hospitals, some of which – including St Bartholomew's in London – Henry had to reopen

*Like other King's Schools, this one at Canterbury was endowed by Henry VIII, and occupies some of the buildings from the monastery he had dissolved.*

(there is a statue of him over the main entrance, as benefactor, although the place would have been there anyway if he had not interfered).

Schools, however, were a different matter; Henry positively wanted to help, and to broaden education to include the masses, since his right-hand man Archbishop Thomas Cranmer said that the aristocratic children 'become unapt to learn and very dolts, as I myself have seen no small number of them very dull and without all manner of capacity', while 'the poor man's son by painstaking for the most part will be learned'. So Henry founded a string of new schools across the country, each called The King's School. Many still used the same old monastic buildings, but Henry typically paid for the education of fifty King's Scholars. Among those who benefited from this improved education system were Christopher Marlowe, the playwright, William Harvey, who discovered the circulation of the blood (page 163), and Isaac Newton, who went to the King's School in Grantham.

When Henry died in 1547, his son Edward succeeded him. The first Protestant monarch, young Edward was clever and well educated, and continued the process of founding new schools, which were called Edward VI Schools. Unfortunately, though, he was continually ill and died, probably of tuberculosis, when he was only fifteen.

# Revolution in physics

The expansion of education was one of the causes of the challenge to existing dogma. Instead of just accepting the wisdom of the ancients, people across Europe were beginning to think for themselves, and to ask difficult questions; the intellectual revolution was gathering speed. A good example was how people thought about the behaviour of inanimate objects.

When considering how and why things behave as they do, Aristotle had been dogmatic, and his logic was with hindsight rather questionable. He asserted that every object wants to return to where it belongs – to its natural place, which depended on how it was made up of the elements earth, air, fire, and water. So if you drop a stone (made mainly of earth) it will fall to where it belongs, on the earth. Pour water from a jug and it will return to the river or the sea from where it came. The natural place of any sort of air – bubbles and so on – was in the blanket of air around the Earth where it belongs, and flames burn upwards because the element fire belongs overhead, towards the sun.

He further asserted that big stones fall faster than small ones, because the heavier one is 'more earthy', and that if one stone weighs twice as much as another it falls twice as fast. Further, he said that any stone falls with increasing speed, because as it approaches the Earth, it falls more jubilantly. There were three types of motion: natural motion was caused by gravity; forced motion was caused by outside forces, such as the action of a bowstring on an arrow; and voluntary motion was caused by the will of animals.

Rather surprisingly, no one seems to have questioned Aristotle for more than 1,900 years; people just assumed he was right. The first person to test his assertions by experiment was probably the Dutchman Simon Stevin, who apparently in 1587 dropped objects of various weights from the top of the leaning tower of the Oudekirk, the old church near the marketplace in Delft. If you lean far out through the stone 'battlements' you can just about drop things into the canal below, which is a good idea, for two reasons. First, it reduces the chance of hitting anyone, and second, the splash gives a good visual clue of the moment of landing. What he was trying to show was that Aristotle was wrong; heavy things do not necessarily fall faster than light ones. Drop a brick and a half brick, and they fall at the same speed: from the tower of the Oudekirk they would make simultaneous splashes in the canal.

Unfortunately, although Stevin was a brilliant mathematician and engineer, and introduced several important ideas, he wrote only in Dutch, and as a result never became well known; indeed, his work has often been ignored. That is one reason why the dropping experiments we hear about are those said to have been done by Galileo from the Leaning Tower of Pisa.

Galileo Galilei was the archetypal Renaissance scientist; he was critical of all established ideas, forceful in promoting new ones, interested in a wide range of things, arrogant, self-opinionated, and keen to be famous. He was born on 15 February 1564, the same year in which Michelangelo and Vesalius died and William Shakespeare was born. His father Vincenzo was a wool merchant and musician, and wanted him to become a doctor, but Galileo hated the old-fashioned teaching according to Galen, and much preferred mathematics, even though that was not meant to be part of his course. After several years of struggling he finally managed, in 1589, to get a job as Professor of Mathematics at Pisa University, where he soon started attacking the Aristotelian dogma taught by the other professors.

The story goes that he announced to the community that he was going to disprove Aristotle's assertions by climbing the Leaning Tower of Pisa and dropping a variety of balls of the same size but of various weights, because they were made of different materials – lead, brass, wood and so on. Hundreds of people are supposed to have turned up to watch this brash young man trying to overthrow the established 'truth'. Aristotle would have predicted that if two balls were dropped together, and one ball weighed twice as much as the other, then it would fall twice as fast. Galileo said that on the contrary, they would fall at almost the same speed – exactly the same speed if there was no air resistance – and hit the ground at almost the same instant.

He may really have dropped light and heavy balls from the top level of the tower – a 50-metre drop on to grass – but the chances are he did no such thing, because it is a poor experiment. You cannot be both at the top, to make sure you drop the objects simultaneously, and at the bottom, to see them land. And they fall so quickly – after 50 metres they would be falling at about 30 metres per second – that it is extremely difficult, even if you are watching carefully, to see which of two things hits the ground first.

To substantiate his idea, it is much more likely that he did a 'thought experiment', which is actually more powerful. Imagine dropping two identical bricks from the top of the tower. They must fall at the same speed. Now imagine that one of them has a crack right through it, and that half-way down it separates into two half bricks. Will those half bricks suddenly slow down and fall at half speed? And what if they are taped tightly together, so that they are just like the original brick? Whether the brick falls in one piece or two cannot make any difference; the speed must stay the same. This is the case for all falling bodies; provided they have roughly the same density and shape, they will all fall at the same speed.

Objects with low density, such as feathers, or objects that are spread out, such as flat sheets of paper, will be much more held back by air resistance or drag, and so fall more slowly, but as long as air resistance can be minimized or kept constant by using objects of similar shape, then big ones and small ones will fall at the same speed.

Galileo was keen to investigate the science of falling, and he devised a cunning way to slow the process down, so that he could observe it better. He rolled balls down a shallow slope, which is just like making them fall slowly. He did not have a good clock, but he checked the timing by hanging bells at intervals above his sloping track so that the ball would make each

*Galileo Galilei, the archetypal Renaissance scientist.*

one 'ting' in turn. Then by listening to the tings he could hear when the ball reached each of the bells.

When he placed the bells at equal intervals down the track – say at 1 cubit, 2 cubits, 3 cubits, 4 cubits and so on (a cubit is the distance from your elbow to your fingertips) – he could hear that the ball was hitting them at shorter and shorter intervals: in other words, it was accelerating, as Aristotle had said. By trial and error, Galileo found that the ball hit the bells at equal intervals of time when the bells were placed at 1 cubit, 4 cubits, 9 cubits, 16 cubits, 25 cubits and so on – in other words, at distances of $1^2, 2^2, 3^2, 4^2$ and $5^2$: the squares of the numbers. He explained that the velocity of the ball at any moment was proportional not to the distance it had travelled but to the time it had travelled.

Galileo was unable to go on from these observations to work out the laws of motion; that was left to Isaac Newton some seventy years later. However, he was one of the first to do experiments in order to try to understand the behaviour of things, and he initiated a trend that eventually became modern science.

*Replica of the pendulum clock that Galileo designed but never completed before he died in 1642.*

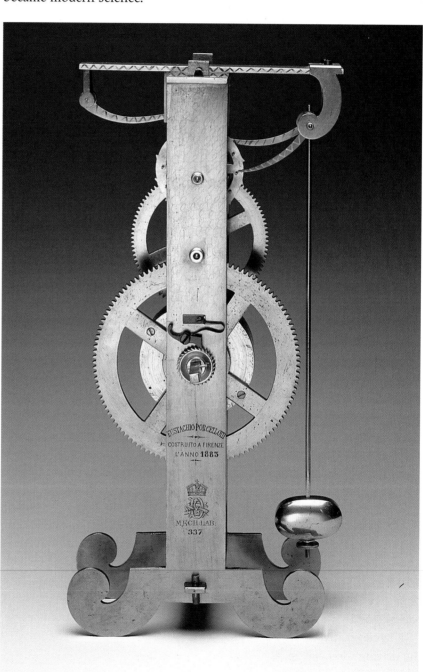

## *The swinging lamp*

One of the other critical observations that Galileo made way back in 1583, while he was still a medical student at Pisa, was to do with the regularity of a pendulum. The story goes that one day he was sitting in the beautiful duomo or cathedral, only a stone's throw from the leaning tower in the Piazza dei Miracoli, and listening to a boring sermon. Right in the middle of the cathedral hung a huge ornate bronze lamp – there is one there today, but not the same one. In the gentle breeze, the lamp began to swing, at first just a little and later in much wider sweeps. Galileo wondered idly how long the lamp took to swing to and fro, and he timed it, using his pulse as a timekeeper. He soon realized that the pendulum took exactly the same time for each swing, regardless of whether it was swinging only a few inches or several feet.

Later he experimented with various kinds of pendulum, and found that the time of the swing depends not on the size of the swing, nor on the weight of the bob, but only on the length of the string. Because the time of each swing seemed so regular, he proposed that a pendulum might make a basic timekeeper, and he built a little pendulum device to measure a person's pulse rate; the length of the string can be varied until the swing just coincides with the pulse, and the rate can be read off a scale. The university medical department claimed the credit for this device, and used it for many years.

Galileo realized that the idea of using a pendulum as a precise timekeeper could have a much wider application; it might form the basis of an accurate clock. Sadly, he never got around to building one before he died, and the idea was not taken up until 1656, when the Dutch scientist Christiaan Huygens built the first pendulum clocks (page 211). Even more sadly, the bronze lamp that now hangs in the duomo is fastened up, and cannot swing in the breeze; so it is no longer possible to repeat this elegant experiment of the great Galileo Galilei.

# Revolution in time-keeping

By the end of the sixteenth century, timekeeping was becoming more precise. The Egyptian water clocks that had been around for more than a thousand years had been superseded by the first mechanical clocks, which were driven by falling weights. The weight provided a constant pull on a piece of string, and this turned a pointer – the hand of the clock – round a dial.

The clever invention was the escapement mechanism, which couples the driving force to the dial and prevents the weight from falling at ever-increasing speed; if there were no escapement mechanism the power

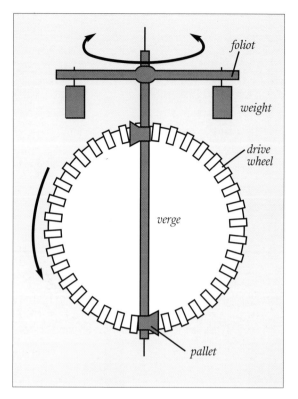

*The verge-and-foliot escape mechanism, used in clocks before 1656.*

source would drive the hands round far too quickly. The first type of escapement was the verge-and-foliot system. The vertical rod or verge has two arms or pallets, which are batted alternately by the pegs on the drive wheel. The top arm is batted to the left, the bottom to the right, and so the verge oscillates to and fro, and each contact stops the drive wheel; so it never has a chance to accelerate.

Immediately after each contact, the drive wheel escapes from the blocking arm; hence the name 'escapement mechanism'. The time taken for the verge to swing through 90 degrees and release the drive wheel can be adjusted by moving the weights in or out along the arms of the foliot: the further out the weights, the slower the movement, because the foliot then has a higher moment of rotational inertia.

Watches presented a different set of problems. Clearly, a portable timepiece could not be powered by falling weights, so the first thing needed was a power spring. This was invented around 1500, along with the fusee, a cone on which the string was wound, so that when it was fully wound up it pulled on the narrow end, but as it unwound it steadily pulled on wider and wider sections of the cone, so that the resulting force on the axle remained roughly constant. You can see a fusee on the top left of the jacket of this book.

There are rumours that Henry VIII had a pocket watch, and the back of it may be what looks like a medallion in some portraits. However, there is no record of such a watch in the inventories of his possessions, and in any case the technology then was not good enough for accurate watches; they were essentially prestigious pieces of jewellery – until religion caused another upheaval. In France from about 1560 the steadily increasing number of Protestants, called Huguenots, were persecuted in a succession of bloody wars and treacherous acts by the ruling Catholics. Thus a treaty giving them some freedom was signed by Catherine de Medici and Charles IX in 1570, but on 24 August 1572 thousands were slaughtered in the St Bartholomew's Day Massacre in Paris. Great numbers of Huguenots fled to neighbouring countries, and those who came to England brought with them certain knowledge and skills that were to have a lasting effect on British craftsmanship.

In particular, the Huguenots had developed some high-tech watches that were much better than any in Britain, and soon the knowledge spread, so that by 1600 there had emerged a definitive English style of watch-making,

and London became the horological centre of the world. Tudor watches, designed to be worn on a chain round the neck, had only one hand, and were hopelessly inaccurate. But they were nonetheless the most complicated machines of the age; these European asylum-seekers had brought to England their skills at the cutting edge of technology.

Many of these Huguenots settled originally in Canterbury, especially in cheap houses along the River Stour. The weavers among them needed water from the river for treating their cloth; among the most famous of the families were the Courtaulds, which is still a well-known name today. So many French people settled in the town that Queen Elizabeth gave permission for a French service to be held in the crypt of the cathedral – and it is still held there every Sunday.

So what the Tudors did for us in all these wide-ranging disciplines was, essentially, to allow new thinking to flourish and spread in the form of the printed word in English: a new Bible, new mathematics, new astronomy, new physics, new anatomy, new medicine, and new imported technology. Britain gradually became the leader in all these fields, and as a result would emerge as one of the most scientifically advanced nations in the world.

*A Tudor watch, designed to be hung on a chain round the neck. There is no minute hand because the timekeeping was so imprecise.*

Being the Second Chapter

# THE WAR MACHINE

'I know I have the body of a weak and feeble woman, but I have the heart and stomach of a King, and of a King of England too, and think foul scorn that Parma or Spain or any prince of Europe should dare to invade the borders of my realm...'

QUEEN ELIZABETH I

enry VII managed to keep his country more or less peaceful during his reign – although he had to contend with a couple of pretenders to the throne. By the time he died, relations with the other European powers were reasonably cordial; and England was now one of the smaller players in the field. All this was to change when his son, Henry VIII, burst on to the scene. A very different man from his shrewd, cautious father, the ebullient new king dreamed of power and glory – to be achieved by military conquest. His ambitious campaigns introduced an arms race: the Tudor period laid the foundations for Britain's later supremacy on land and sea, with significant developments in warships and artillery technology, and also in the more shadowy world of espionage.

## Henry the warmonger

Henry VIII became king in 1509, just before his eighteenth birthday. In sharp contrast to the later image of him as an enormously fat old man, in his youth he was athletic, tall, and strong. He gloried in all manner of physical pursuits, especially hunting and jousting. As well as its intrinsic pleasure, such vigorous sport kept the king and his courtiers fit and healthy – and part of this endeavour was in preparation for war. A Venetian diplomat wrote, 'The new king is eighteen years old, a worthy king and most hostile to France… it is thought he will indubitably invade France.' And so he did.

By the end of June 1513 Henry had assembled a great fleet of 300 ships in the Channel at Dover, packed with troops who were protected by 12,000 newly-made suits of armour. To feed the men, 25,000 oxen had been slaughtered and salted down. Henry himself carried the most fabulous equipment: gold trappings for his horse, gold buttons for his doublets, and a silver crossbow in a silver-gilt case. He was rowed out to his ship wearing over his armour a white tunic with a red cross on the chest – the outfit of a crusading knight. He saw himself as part crusader, part Henry V, who almost a hundred years before had crossed the Channel and flattened the French army at Agincourt. He felt it was his duty to seek fame by military skill; he wanted to be chivalrous, brave – and victorious.

Furthermore, the French King, Louis XII, had laid siege to the Pope's forces at Bologna, and Henry knew he must be in the right to be fighting against the enemy of the Pope (this was long before he himself became just such an enemy). He even acquired the papal indulgence for his soldiers, which meant that if they died in battle their sins would be forgiven. Henry first laid siege to Thérouanne, bombarding it while he built a great town of tents outside; he himself had eleven tents and a wooden house to retire to. He did not hide, however, but rode about in the open in his crusader's outfit and 'a red shaggy hat with many red feathers'.

'The new king is eighteen years old, a worthy king and most hostile to France... it is thought he will indubitably invade France.'

VENETIAN DIPLOMAT

Two thousand French knights came riding up to try and relieve the siege, and were successfully attacked by eleven hundred English. Henry was prevented from leading the charge, but joined it soon afterwards and exulted in pursuing a defeated enemy. This rather grandly named 'Battle of the Spurs' was his finest victory against the French. He eventually took Thérouanne, Tournai and five other substantial towns, before calling off the campaign in October. Escorted back across the Channel by his flagship the *Mary Rose*, Henry became famous; tales of his bravery, skill and chivalry were told across the country. He had become Great Harry.

However, while he was away he had left his pregnant wife Catherine in charge of the country, and when James IV of Scotland invaded with 20,000 men she had to organize an army in defence, under the command of the

*King Henry VIII was an enthusiastic sportsman and insisted that all young men should practise archery.*

*Catherine of Aragon was married to Henry VIII in 1509 and remained faithful for more than 20 years – but failed to produce a son and heir.*

seventy-year-old Earl of Surrey. The result was the Battle of Flodden, on 9 September, in which the English archers played a decisive role – this may have been the last battle won by the longbow. The Scots were utterly defeated, and their king was killed. This was of far greater military significance than anything Henry had achieved in France. Moreover, Henry's campaign had cost the country nearly a million pounds, all the money he had inherited from his thrifty father.

Where did all this money go? Apart from Henry's own lavish accoutrements, he had ordered forty-eight heavy guns from the finest gun-makers in Flanders, along with thousands of handguns – even though these were as yet no match for the longbows – and vast quantities of gunpowder.

Above all, he ordered the building of the *Mary Rose* and other warships. Henry unleashed further invasions of France in 1523 and 1544, while in 1539 the French almost invaded England. None of these campaigns was decisive, except in their cost.

When Henry needed to replenish his Treasury, his eye fell on the wealthiest organization in the land: the Church – the Catholic Church, that is, from which Henry had by 1533 so decisively split (see page 26). Henry, as Supreme Head of the Church in England, decided to dissolve the monasteries and appropriate their assets. There were hundreds of these Roman Catholic enclaves, and in 1535 and 1536 Henry simply had them demolished, and confiscated their land and their possessions, keeping the money and selling the land to wealthy Protestants. There were several major consequences of this earth-shaking process. Many beautiful buildings were destroyed; some of the ruins are still visible today. The Catholics received a series of terrible blows: personal, financial and spiritual. Meanwhile, the sale of vast tracts of territory completely shifted the balance of land ownership in the country. However, in the short term, the dissolution of the monasteries solved Henry's financial problems.

# The Royal Navy

Henry VIII can be seen as 'father of the Royal Navy', because he was the first king to keep a standing navy. Before this, the king would put a naval fleet together when he wanted to go to war, and then when the war was over he either sold it or scrapped it. But in 1514, after his first invasion of France, Henry decided to 'mothball' his warships, and maintain them ready for use; he was fairly sure he would be invading France again before too long. He realized that both for offensive action against France and for the defence of the island in general he would need long-term control of the seas – a goal that would be vindicated in spectacular fashion when his daughter Elizabeth was threatened by the Spanish Armada in 1588 (see page 70).

Accordingly, Henry started putting together a team of naval administrators; in that same year, 1514, he gave a Royal Charter to 'The Master, Wardens, and Assistants of the Guild, Fraternity, or Brotherhood of the Most Glorious and Undividable Trinity and of St Clement in the Parish of Deptford Strond in the County of Kent'. This ponderously named body became known as Trinity House; it now occupies a fine building by the Tower of London, and is responsible among other things for all the lighthouses around the country. At the end of his reign in 1545 Henry created a new department of state called the Council for Marine Causes, which was responsible for the king's ships and dockyards.

Henry was also responsible for the creation of a new kind of ship: the battleship – powered by sail and heavily armed. No longer content to adapt

*The* Matthew, *a replica of John Cabot's ship. The original was built in 1497 with a fine carvel hull.*

merchant vessels for use as armoured troop carriers, he wanted to be able to engage in naval warfare, and so he ordered his shipwrights to design and build boats specifically with war in mind. In particular, they constructed boats to carry and fire the most up-to-date military equipment – cannons (see page 61). At first, cannons were simply mounted at either end of the decks of existing vessels, but this caused problems. Because of their weight, the cannons tended to unbalance the ship, and few guns could be carried so high above the water line. Worse, the gun crews could not fire broadsides, because the guns had to point fore and aft. The solution to these difficulties was a vital innovation: the gun-port.

The shipwrights had already embarked on a new way of making the hull; instead of clinker-built construction, with overlapping planks, they introduced the carvel hull, in which the planks were butted together side by side. Carvel had also come to mean a type of ship, including the *Matthew* (see page 90). Gun-ports needed to be watertight in case of rough weather and big waves, and watertight gun-ports were easier to construct in a carvel-built hull.

Side ports (*porte* is French for door) had been used for some time in loading and unloading the ships. They were essentially holes cut in the side of the ship, which meant the sailors could unload cargo easily when the boat came into port whatever the tide and hence whatever the height of the decks. At some point, these side ports were adapted for guns. The first references to gun-ports are from around 1510, and they may have been an English invention – possibly even by the king himself.

There is some debate as to which boat was the first to be built specifically to carry heavy guns. Conventionally, the *Mary Rose* has been seen as an experimental ship, but some historians believe that the Scottish king, James IV, led the way with the construction of his ship, the *Michael*, in 1506. In any case, the *Mary Rose* and the *Peter Pomegranate* were both built in 1510, at a total cost of about £1,000. Henry allegedly named the *Mary Rose* after his sister Mary and the Tudor Rose, and he spent a further £200 on flags and banners to decorate her as his flagship.

The *Mary Rose* was one of the biggest and best-armed ships of her day and a prime example of the new-style English Navy. She was some 40 metres long, displaced some 700 tons, had a crew of between 400 and 500 men, and outsailed all the other ships in the fleet. She had cannons on three decks, fired through three rows of gun-ports. Each of the 32-pounders on the lowest deck was manned by a crew of five, including the port-man, whose job it was to open the gun-port before firing, and close it again immediately afterwards.

*Gun-ports on* HMS Victory; *their design seems to have changed little since the first Tudor warships of around 1510.*

Before the arrival of heavy cannons on board, the principle of naval warfare was not to blow the enemy ship out of the water, but instead come alongside and grapple with her, so that armed men could fight their way on board, attack the crew, and ultimately take control of the vessel – like taking a tract of land in a land battle. Ships were very expensive to build and, if there was any chance of capturing one from the enemy, that was much better than sinking her.

Eventually, by the middle of the sixteenth century, side-firing of guns became standard, and tactics began to change. The English in particular developed the idea of holing and sinking ships rather than taking them. Before then, the guns were fired when the boat was on the 'up roll' so that the cannonballs or shot were aimed upwards, either towards the rigging – to bring down the sails and disable the ship – or at the soldiers on board. However, the English began firing on the 'down roll', aiming to put a hole in the side of the enemy vessel near the waterline.

One crucial innovation that does seem to have been English was the dry dock. Originally, boats were built in dockyards that were blocked with a mud dam at one end. When the shipwrights had finished working on the boat, the dam had to be dug out in order to let the water in and float the ship. This digging took many men several weeks. With the early-sixteenth-century construction at Deptford of the first real dry dock – a dock with watertight gates – this problem was solved, since the gates could be opened and closed at will. No foreign naval power acquired a dry dock until 1666, and by then the English had a technological lead which they never lost.

Some people have claimed that shipbuilding made heavy demands on the forests, and that timber was in short supply, but this turns out to be wrong. Apart from the masts, warships were made entirely from oak, and the shipwrights looked particularly for oak from parks and hedges, rather than from forests, because they had curved branches that were most useful for the curved timbers needed in the ships. There was plenty of oak about, and the Navy never seems to have bothered even to look in the New Forest or the Forest of Dean until the late 1600s. Even for Nelson's navy, sixty times the size of Elizabeth's, there was plenty of available oak.

# Defending the realm

Apart from the ever-present threat from France, there was another major element that affected Henry's military and strategic planning. His rift with Rome had naturally outraged the Pope, who in 1533 excommunicated the king – the ultimate punishment, casting him out from the Catholic Church. In addition, however, the Pope encouraged the rest of Catholic Europe to combine forces against this heretic who had dared to start his own church. In particular, he urged Spain and France to co-operate in a crusade against England. So by the early 1540s Henry, at war with all his neighbours, was seriously worried about an invasion by the French, perhaps combined with the Spanish, and he began to commit more and more resources to the defence of the country.

Among other projects, he constructed twenty-six new castles to make a chain of coastal defences all along the south coast from Tilbury to Cornwall. Tudor castles were the first to be designed with cannon emplacements.

Before, English castles had had keyhole openings in their walls for archers and for hand guns but not for cannons. One of these new castles, at Southsea, was built in 1544 in less than six months, and it was specially important because of its key position alongside the deep water channel that runs into Portsmouth, which was where most of Henry's new shipbuilding took place. In other words, Southsea Castle had to defend the naval dockyards, and so it was designed with the latest in Tudor technology.

Based on an innovative Italian design, Southsea was the first castle of its kind to be built in England. What made it special was its shape. Earlier fortifications, such as Windsor Castle, had rounded towers, but Southsea was built with straight walls that met to form pointed bastions. These pointed fronts were more easily defended. At the foot of the rounded towers there

*Southsea Castle, built in 1544, was King Henry VIII's first castle built with a square and pointed shape, which left no blind spots for enemies to hide in.*

*This picture shows Henry arriving (bottom right) in time to see his beloved flagship* Mary Rose *sinking in the Solent (top).*

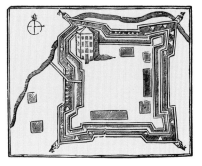

*Plan of a typical castle from the period showing pointed bastions that were more easily defended.*

*The remains of the* Mary Rose, *pulled up in 1981 after 536 years in the mud of the Solent, and now being continuously sprayed with polyethylene glycol to preserve the wood.*

had always been a blind spot where the enemy were safe from defensive fire; the only way to attack them was to drop something on them from above. But cannons could fire all the way along the angled bastions – towards the pointed tips – leaving nowhere for the attacking enemy to hide.

According to a letter written by the Governor of Portsmouth, Sir Anthony Knyvet, who supervised its construction, Henry VIII designed Southsea Castle himself. Knyvet wrote that he hoped the King would be pleased with the castle 'which was of his Majesty's own device'. Furthermore, legend has it that Henry VIII watched the *Mary Rose* sink from the battlements of Southsea Castle.

This pride of the fleet met her end in 1545, when the threatened invasion almost happened: a huge French fleet of 200 ships sailed into the Solent, intending to invade the Isle of Wight and destroy the English fleet, of which there were only about eighty ships in Portsmouth. On 19 July, the third day of the engagement, the *Mary Rose* sank in what some reports say was a flat calm, and all her crew were drowned. No one knows why she sank, but the bottom row of gun-ports was only about a metre above the waterline, and it seems likely that after firing the cannons the gun-port men did not have enough time to close the ports. The ship heeled over – perhaps there was a freak gust of wind – and the sea poured in. Then the heavy guns provided plenty of weight to take her to the bottom.

## Cannons

After building so many new castles, Henry had to equip them all with the latest in armaments, and this brought about a major new industry in Britain: cannon-making. Cannons had been fired at the Battle of Crécy in 1346, but they were not much use then; the longbow was not completely superseded until the end of the sixteenth century. Early cannons were really just buckets with a touch-hole in the bottom, but then the makers began constructing them from wrought iron bars, bound in a cylinder round a wooden former, and fixed together before the former was removed. The next stage was to cast cannons in one piece by pouring molten metal into moulds, starting with bronze.

Casting cannons was a highly skilled process. First, the cannon-maker would carve a replica of the cannon from wood. Then he would build a mould of clay around it, after smearing the wood with clay, grease, dung or ashes to prevent the clay from sticking. He would reinforce the clay either with iron rods, or by binding with wire, and leave it to dry; then he would carefully remove the wooden former and bake the clay mould. He would then lower it into a pit, put a core inside to make the hollow inside the barrel, and pour molten metal into it, straight from the furnace. When the metal had cooled he broke open the mould, and bored out any irregularities inside the barrel.

*Bronze cannons were made before Tudor times, but were expensive and liable to buckle if fired too many times in quick succession.*

In the first twenty years of his reign, Henry imported from abroad 140 bronze cannons, including huge bombards that came to be called the Twelve Apostles. Many of these guns came from the famed Flemish gun-maker Poppenruyter. Not only were they expensive, producing a severe drain on his cash reserves, but in the early 1530s, after his schism from Rome, the supply of guns from Europe was cut off. So he started the Royal Armouries, importing French and German craftsmen to make his bronze cannons. However, the copper and the tin needed to make the bronze still had to be expensively imported. The next big step forward had to wait for improvements in the smelting of iron, in the Weald of Kent.

Bronze had been known and used for thousands of years – the Colossus of Rhodes was a bronze statue of Apollo, 25 metres high, and erected in about 280 BC. Chemically, bronze is an alloy – a mixture – of copper and tin, and it melts at around 900°C, depending on the composition. Iron melts at above 1500 °C; this much higher temperature makes it enormously

*Iron foundry from* De Re Metallica *(1556), showing the furnace at the top and the red-hot iron 'bloom' being wrought (i.e. hammered into shape) by a blacksmith in the centre.*

more difficult to smelt, or extract from its ore. Before about 1500, iron was made in what was called a bloomery, where a small quantity of iron ore was smelted with charcoal in a furnace to give a spongy mass or 'bloom' of iron, which was taken out still red hot, and hammered into the required shape. This hammering made it wrought iron, which is tough and fibrous.

The next step was the blast furnace: a big water-wheel drove mechanical bellows to blast air into the furnace and generate a temperature above 1500°C. Inside, a large quantity of iron ore was smelted with charcoal to make molten iron, with slag floating on top. Blast furnaces had to be built where both iron ore and wood were available beside a stream – and the Weald of Sussex and Kent provided all these facilities. In the blast furnace the molten iron absorbed some of the carbon, so that the molten metal that was tapped off from the bottom of the furnace contained a few per cent of carbon. This was cast iron, harder than wrought iron, and strong in compression, but rather brittle. However, the molten iron could be run directly into moulds, to form complex objects in a single operation, and much more iron could be produced in the blast furnace than in a bloomery.

In 1556 a book called *De Re Metallica* was published, the first ever systematic account of mining and minerals, written by a German, Georg Bauer, who called himself by the Latin name Agricola. This book, illustrated with many woodcuts showing industrial processes, remained a basic text for at least a hundred years, and provided a dramatic illustration of the power of printing in the spreading of knowledge. The description of the iron-smelting process in *De Re Metallica* shows that in Germany by 1556 the quantities were still small, and the blast furnaces fairly primitive:

'The hearth is three and a half feet high, and five feet long and wide; in the centre is a crucible a foot deep and one and a half feet wide... The master first throws charcoal into the crucible, and sprinkles over it an iron shovel-ful of crushed iron ore mixed with unslaked lime. Then he repeatedly throws on charcoal and sprinkles it with ore, and continues this until he has slowly built up a heap; it melts when the charcoal has been kindled and the fire violently stimulated by the blast of the bellows... He is able to complete this work sometimes in eight hours, sometimes in ten, and again sometimes in twelve.'

English technology seems to have been creeping ahead during this period. The first English blast furnace was built in the Weald in 1496, and know-how gradually developed. Some thirty years later, an illiterate farmer's son called Ralph Hogge became the King's 'gunstonemaker'. His job was to carve stone cannonballs, but he became a skilled blast-furnace operator, with a furnace as big as a house. By the early 1540s Hogge was casting spherical iron cannonballs, and in 1543 he was able to cast the first iron cannon. According to legend:

> Master Hugget and his man John
> They did make the first cannon.

At this time most of the cannons in Germany and Flanders were still cast in bronze – easier to handle as it melts at a much lower temperature. Moreover, while iron-founding was still developing there, the furnaces tended not to be hot enough, and so the iron they produced was inferior; cannons cast from it were liable to burst. However, Hogge and his successors were able to make iron cannons that were both better and cheaper than the bronze cannons from the Continent, and before the end of the century developed a considerable export market for their guns. Because this iron-casting technology came together in the Weald, England had almost a monopoly of cast-iron cannons from the time of Henry VIII for nearly a hundred years.

Cast-iron cannonballs, were a considerable improvement on stone ones; it was easier to make them accurately spherical, so that they flew straighter; they were denser, so that they were more penetrating; and in principle they could

*Above: Nicholas Hall, Keeper of Artillery, with a 32-pound cast-iron cannonball.*

*Below: The first cast-iron cannon was made in the Weald of Kent in 1543; this one, also cast in iron, was made for the BBC.*

be made to a fairly standard size, so the balls fitted better in the barrels of the guns, and had a higher muzzle velocity. Not that the inside of a cast-iron gun barrel was ever smooth; there were always lumps and bumps, and to be sure of getting out, the balls had to be made about 10 per cent narrower than the barrels. This meant that the guns were inefficient, because a good deal of the blast of the gunpowder was wasted around the sides of the ball. Also, the ball bounced to and fro inside the barrel and came out in a fairly unpredictable direction; sixteenth-century cannons were never accurate at ranges of more than about 100 yards. The first man to make really smooth cylindrical barrels was John 'Iron-mad' Wilkinson, and that was not until 1775.

## Gunpowder

However good the cannons, they were useless without gunpowder, which was invented in China at least a thousand years ago. The first published information about it in Britain was written in the thirteenth century by a monk called Roger Bacon, who said, 'The sound of thunder may be artificially produced in the air,' and gave the formula for gunpowder in the form of an anagram in Latin, which apparently lay undeciphered for 650 years. His formula, roughly 30 per cent each of charcoal and sulphur and 40 per cent of saltpetre, is in fact too low in saltpetre to be explosive. It fizzes and pops, but would hardly propel a cannonball. Gradually, the practitioners realized that they needed more saltpetre, and increased the proportion to about 75 per cent, with 12 per cent each of charcoal and sulphur. Those proportions are still used today.

Saltpetre, or potassium nitrate, is the key to good gunpowder, because it provides the oxygen for the burning of the other chemicals. Both charcoal (carbon) and sulphur react with this oxygen to make gases – carbon dioxide and sulphur dioxide. Therefore when solid gunpowder is ignited, in a complicated reaction it rapidly produces a huge volume of hot gas, and if it is confined in a box or a bottle or a bomb or a cannon this makes an explosion.

Saltpetre is formed naturally when organic matter decomposes – in compost heaps, for example, although rain will wash it out. Saltpetre can build up on walls in sheds and stables, where it is protected from rain, and in countries where there are long dry seasons. India used to export many thousands of tons of saltpetre every year from the sewage-soaked earth round villages, and the mud walls of huts. To make effective gunpowder the saltpetre must be pure, and it can be purified to some extent by careful crystallization. Here is part of a recipe, translated from the Latin, from *De Re Metallica* (see page 62):

> Saltpetre is made from a dry, slightly fatty earth, which, if it is retained
> for a while in the mouth, has an acrid and salty taste. This earth,

*together with a powder, are alternately put into a vat in layers a palm deep. The powder consists of two parts of unslaked lime and three parts of ashes of oak… water is poured in until it is full. As the water percolates through the material it dissolves the saltpetre; then, the plug being pulled out from the vat, the solution is drained into a tub and ladled out into small vats… poured into the rectangular copper cauldron and evaporated to one half by boiling…*

…and so on for two more pages. This operation clearly needed skill and experience. In 1558 Queen Elizabeth gave an eleven-year monopoly for the manufacture of saltpetre to George Evelyn, grandfather of the diarist John Evelyn (who was to become a successful gunpowder manufacturer himself). Then in 1561 she bought, for the princely sum of £300, what was alleged to be a better recipe from a German captain called Gerrard Honrick:

*First take black earth, the blacker the better.*
*Then a good dose of urine from someone who drinks strong beer or wine.*
*A nice load of fresh dung from a horse who likes to eat oats.*
*Mix it all together with some plaster of Paris.*
*Leave it to fester for a while, turning every week or two.*

In three months this was supposed to generate wonderful saltpetre, after the organic matter had decomposed and the ammonia from the urine had been oxidized to make nitrate. Honrick gave a handy tip to find out when the saltpetre was ready: dip your finger in it, and it should taste salty.

Saltpetre men were even given royal permission to collect dung-soaked earth from houses in London, and householders were forbidden to pave the floors of their stables and pigeon houses to make sure the earth was continually impregnated. In 1625 Charles I was to issue a proclamation to the effect that all 'his loving subjects… shall carefully and constantly keep and preserve all the urine of man during the whole year, and all the stale of beasts which they can save and gather together while their beasts are in their stables and stalls…' However, this was asking a good deal of his subjects, and may even have been one of the factors that contributed to his downfall a few years later. From about 1630 much of the saltpetre was imported from India, which allowed the urine to be used for the production of alum (see page 200).

Once the saltpetre is dry and fairly pure, it should be combined intimately with the charcoal and the sulphur. In the early days the powders were just mixed, and that would give a mixture that would burn fiercely. For a powerful explosive mixture, however, the ingredients have to be ground together, so that the powders are extremely fine and completely mixed. This grinding was a dangerous process, since any spark could produce devastation. Generally, therefore, the saltpetre was transported

'His loving subjects… shall carefully and constantly keep and preserve all the urine of man during the whole year, and all the stale of beasts which they can save and gather together while their beasts are in their stables and stalls'

PROCLAMATION BY CHARLES I

*A wheel-lock Tudor pistol, with elegant carved hartshorn decoration.*

separately from the other ingredients, and the grinding was done at the last moment – on the battlefield – by the 'fireworker', who was always a highly respected member of the team. He usually made a mortar by burning out the trunk of a tree (preferably alder or willow) and then grinding the mixture in the mortar with a pestle. After a victory in battle, the fireworker would traditionally celebrate by setting off any remaining gunpowder in a display of fireworks.

On the battlefield, cannons had to be fired by holding a smouldering wick to the pan, where with luck it would light the gunpowder in the touch-hole and so ignite the charge inside. The system was unreliable, though; the wick was liable to go out in wet weather. The match-lock pistol, which needed two hands to fire, was little better. Henry VIII, however, had a pistol with a wheel-lock, which he proudly used when hunting. The wheel-lock was apparently designed by Leonardo da Vinci around 1500. A chain wound around a spring-loaded steel wheel makes it spin rapidly, rubbing against a piece of fool's gold. The friction generates a spark, which ignites the gunpowder. When this worked it was superb, and was not affected by rain. However, the wheel lock was expensive to make, and tended to get gummed up on the battlefield. Nevertheless, it was the precursor of the self-igniting gun, and eventually led to the flintlock and then the percussion cap.

# The secret service

International politics were also rather explosive in the sixteenth century, and enormously complicated: who went to war with whom usually depended on which secret alliances or betrothals had just been made – or broken. Knowledge of what was going on in other countries was vital but often woefully inadequate. There were no permanent English embassies abroad except in Paris; so spying was a necessary part of policy-making.

Henry VIII had rather a primitive spy network, but his daughter Elizabeth took espionage to a new level. Her spymasters took pains to recruit highly skilled agents, often young graduates from Oxford and Cambridge – which still today form a hunting-ground for spooks. Many young Elizabethan men were happy to become spies, even though the work was often dangerous and badly paid; at that time there were not many attractive careers for the sons of the educated gentry.

The methods used by plotters to pass secret information centred on elaborate methods of concealment – inside the heels of slippers, for example – and the creation of codes and ciphers. Elizabethan ciphers had two main forms: transposition (when the order of the letters was shuffled) and substitution (when they were replaced with other letters, numbers, symbols and so on). The main problem with both these methods was the time it took to prepare the material, and indeed to decipher it.

*The cardan grille was a perforated piece of leather or similar material that enabled you to read an encoded message: here 'meet at pub at noon'.*

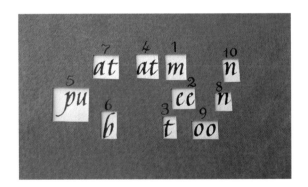

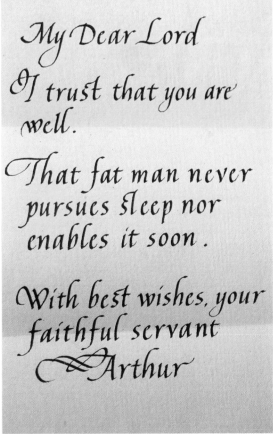

My Dear Lord

I trust that you are well.

That fat man never pursues sleep nor enables it soon.

With best wishes, your faithful servant
Arthur

However, a Milanese doctor and mathematician, Girolamo Cardano, devised a clever piece of technology called the 'cardan grille'. This was a piece of sturdy material pierced with rectangular holes at irregular intervals. These holes were the same height as the text but of various widths, so that when you laid it on top of the paper you could read sometimes whole words, sometimes syllables, or occasionally only single letters. Each hole was randomly numbered, but when you placed the grille on the piece of paper, you could read the message in the number order. The message was disguised by incorporation in an apparently innocuous letter. A decoder with an identical grille placed it on top of this text to read the message in the correct order. This method was widely used for many years but the key problem was the covering message. It was often so clumsy that any censor who intercepted the letter was alerted immediately.

Elizabeth's spy chief was Sir Francis Walsingham, whom she knighted at Windsor Castle on 1 December 1577. He took his secret service so seriously he financed it himself and ran it from his own home. At any one time he had on his books more than a hundred 'correspondents' scattered around Europe. One of his most skilled agents was Thomas Phelippes, a Cambridge MA who was an expert in cryptanalysis and foreign languages. However, many of his agents were simply merchants who sent him any news they had heard.

Walsingham is credited with the biggest coup of the time: the uncovering of what became known as the Babington plot, and the subsequent execution of Mary Queen of Scots. He had always been wary of Mary and the threat she represented to Elizabeth. But he was also aware that Elizabeth would not consider executing her own cousin for anything less than incontrovertible proof of an assassination plot. He suspected that Mary was scheming with various English cronies, but he had to catch her in the act. His chance came in 1586 when he managed to intercept and decode key letters between Mary and the Catholic plotters.

Mary was being held political prisoner by Elizabeth, at a country house called Chartley, and exchanged letters with her co-conspirators using a 'secret' postal system, sealing the messages in watertight bags and hiding them in barrels of beer that were being delivered to the house. Unbeknown to her, Walsingham had actually set up this postal system so that he could track all Mary's correspondence. In 1587 Elizabeth reluctantly signed the death warrant for her cousin, who was executed at the great hall at Fotheringay Castle on 8 February.

## *Torture at the Tower*

The increasing use of spies by all the countries of Europe led to greater reliance on counter-intelligence. Anyone suspected of spying or plotting was likely to be incarcerated in the Tower of London, and there to extract the truth the gaolers usually used torture: a grimly perennial tool in the penal armoury.

The standard piece of equipment for torture was the rack, a large iron bedframe on which the victim was forced to lie. His feet were tied to the bottom of the bed and his hands tied with rope to a winch at the top. Then the winch was used to pull his hands and gradually stretch the body until joints – often the shoulders – dislocated. The tension could be maintained as long as necessary to secure a confession or to uncover the secrets the victim was thought to know.

The rack was highly successful, and was used for example in 1605 to extract a 'confession' from Guy Fawkes after the Gunpowder Plot. However, it had one disadvantage: it was cumbersome and essentially immobile, since it was bolted to the floor in the White Tower, so that it could not be taken to a prisoner's cell. All the victims had to be brought to the rack, and carried back to their cells afterwards, which could be inconvenient.

Henry VIII's Lieutenant of the Tower was Sir Leonard Skevington or Skeffington, and he came up with a new 'engine of torture' that was simple, portable and horribly effective, and became known as 'Skevington's Gyves', or 'the Scavenger's Daughter'. The victim was made to kneel on a plank, each end of which was hinged and attached to a length of iron, curved to make a quarter-circle. This pair of iron 'hoops' was pushed over

*The Skevington irons or 'Scavenger's daughter' – most uncomfortable.*

the small of the victim's back, and connected together by a giant screw; this forced him to crouch uncomfortably down with his shoulders near his knees. This original position was exceedingly uncomfortable, and after a few minutes would produce agonizing cramps. However, the torturer could easily make it much worse, by gradually tightening the screw to bring the ends of the hoops closer together and gradually compressing the victim into a tighter and tighter ball. The screw could be tightened more and more, until blood spurted from the nose, mouth and ears; eventually the vertebrae would dislocate, breastbone and ribs would crack, and the internal organs would be crushed. Even a few minutes of Skevington's Gyves was generally enough to secure the fullest confession – and, purportedly, secure the safety of the realm.

# The Spanish Armada

Soon after the execution of the Catholic Mary Queen of Scots in February 1587, Sir Francis Walsingham received information that Philip II of Spain planned to invade Britain in retaliation. In April, therefore, Francis Drake sailed off and 'singed the King of Spain's beard' by setting fire to the Spanish fleet in Cadiz harbour; he destroyed a hundred ships. Even this was not enough to stop the plans, but it did cause a year's delay, during which the brilliant Spanish admiral, the Marquis of Santa Cruz, died. In 1588 a fleet of 129 Spanish ships sailed up through the Bay of Biscay to the English Channel, carrying 20,000 soldiers and 8,500 sailors. The plan was to take control of the Straits of Dover, collect a waiting army of 35,000 men in Flanders, and then invade Kent, as the Romans had in AD43.

According to legend, Drake was playing bowls on the grass of Plymouth Hoe when the Armada was sighted sailing into the Channel. With supreme confidence, he refused to be diverted until the game was over, and then he sailed off to attack. Meanwhile beacons were lit in a chain along the south coast, so that news of the Armada travelled fast. In fact Drake and the other commanders, Hawkins, Frobisher and their leader the Lord Admiral, Lord Howard of Effingham, had known the Armada was coming and wanted to go and attack it earlier, but were prevented by headwinds. On 20 July the English fleet of some eighty ships beat their way out of Plymouth Sound and during the night sailed past the Spanish to get upwind of them.

Meanwhile, the Queen was magnificent. The Earl of Leicester had assembled 16,000 troops at Tilbury, ready to repel the expected invasion. Ignoring the advice of her ministers, Elizabeth travelled down to encourage them, and in a rousing speech declared, 'I know I have the body of a weak and feeble woman, but I have the heart and stomach of a King, and of a King of England too, and think foul scorn that Parma or Spain or any prince of Europe should dare to invade the borders of my realm...'

The Armada, now under the command of the incompetent Duke of Medina Sidonia, sailed in a tight formation, and the fifty warships – the rest were basically troop-carriers – were armed with heavier guns, but the English ships were more agile and better sailed, and could also fire much more rapidly. This was because the Spanish guns were made of bronze, which after a few shots became hot enough to warp, while iron, because of its higher melting point, did not. So the English guns could be fired again and again with impunity, delivering devastating broadsides intended to disable and if possible sink the enemy ships, rather than employing the old practice of grappling and boarding the vessel.

For nine days the two fleets sailed slowly up the Channel, scrapping here and there. Two Spanish ships were sunk, but little real damage was done,

*Sir Francis Drake, buccaneer and scourge of the Spanish Armada.*

*The English fleet, on the left, engaging the Spanish Armada in the English Channel in July 1588.*

although both sides were running short of supplies of food and ammunition. However, on 27 July the Duke made the fatal error of taking shelter in the harbour at Calais, partly in order to make contact with the waiting army.

The English anchored a mile offshore, and the next night filled eight ships with timber and pitch, set fire to them, and set them drifting with the tide into the Spanish anchorage. Terrified by the threat of being set on fire, the Spanish ships cut their anchor cables, broke formation and fled in confusion.

In the morning the English attacked the scattered ships ferociously off Gravelines, sent the entire Armada in disarray into the North Sea, and pursued them as far as the Firth of Forth. In terrible storms, the Spanish had to make their escape round the north of Scotland, and by the time the fleet limped back into home waters some months later, there were

only fifty-four battered ships and fewer than 10,000 men. The war with Spain rumbled on for another sixteen years, but the rout of the Armada saw the end of serious threats of invasion of Tudor England.

So what the Tudors did for us in the military sphere was to lay the foundations of Britain's strength on land and sea, establishing a navy and an army that would serve the country well in the following centuries of conflict. By the end of their period, England had once more become a major player on the European stage: improved technology gave them a cutting edge in weaponry, complemented by innovative strategic thinking, while their new design of warship was to dominate the seas for 300 years. The seeds were also sown of a system of intelligence-gathering and processing, which would become increasingly sophisticated and underpin the more visible manifestations of power.

*L 'Fleminco.*

Being the Third Chapter

# SEEING THE WORLD

'Some special cunning paintor might be permitted... to take the natural representation of her majesty... after which finished, her majesty will be content that all other painters or gravors shall and may at her pleasures follow the said pattern.'

ELIZABETH I

*Previous spread: a copy of one of John White's watercolours from Roanoake (see page 101).*

he Tudor period brought advances in the technology of drawing that ushered in a dramatic new realism in artistic interpretation. This realism found a natural expression in the production of maps, which for the first time, with the help of newly developed scientific instruments, could be made accurate and reliable. These maps were to prove a vital tool in the burgeoning voyages of exploration around the world in pursuit of trade and new lands to conquer. The world was being seen in different ways, and the images were being indelibly recorded.

# The camera obscura

In the fifth century BC the Chinese philosopher Mo-Ti recorded that by letting light through a tiny hole into a dark room he made an image on the opposite wall – a picture of the world outside that he could see through the hole, but upside-down and reversed from left to right. Aristotle used the same trick to observe an eclipse of the Sun: a clever way to do it, since looking directly at the Sun is dangerous, particularly if you want to do it for a long time. Eighteen hundred years later, the Dutch mathematician Gemma Frisius published a picture of how he had done the same thing to observe the eclipse of the Sun in January 1544.

Around AD 1000 the Arabian scientist Abu Ali Mohamed Ibn al-Hasan Ibn al-Haitham, also known as al-Hazen, had used the same technique to demonstrate that light travels in straight lines. Born in Basra in about AD 965, he was a Muslim – and the Koran encourages believers to observe natural phenomena, 'like the flight of birds and the falling of rain'. In middle age he moved to Egypt, and there in Cairo he showed that with five lamps burning outside the room he could see on the opposite wall inside five images of flames, upside-down. Furthermore, he could remove any of the images from the wall simply by putting his hand between the hole and the image, which meant that the light must be travelling in a straight line from the hole to the image on the wall.

*The camera obscura is a darkened room or box into which light comes from a tiny hole in one wall and casts an image on the opposite wall.*

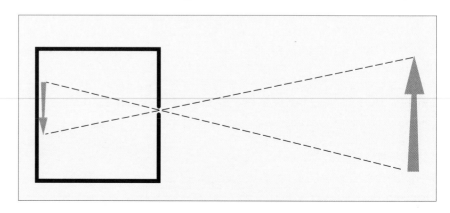

*The camera obscura was a useful new tool for artists.*

This was a remarkable step: al-Hazen gave a clear scientific description of his work, and was actually doing experiments about 600 years before the idea occurred to the natural philosophers of western Europe.

This dark room that was used to see an image came to be called the 'camera obscura'; the Latin expression literally means 'dark room'. You can easily make a small camera obscura using a cardboard box. Remove most of one side of the box, in its place taping a piece of translucent paper (tracing paper, or the inner bag of a cereal packet). Then make a pinhole in the opposite side. A good way to do this is to make a hole big enough to poke a finger through; tape a piece of kitchen foil over the hole, and use a pin to make a clean round hole in the foil. This way you can easily make a new pinhole, larger or smaller as necessary.

On a bright sunny day, put your box down with the hole pointing at a pleasant view, and look at the paper back, covering your head and the back of the box with a thick dark coat or cloth to keep all light off the back of the box; the paper must be lit only from inside the box. Then you should see an image of the view upside-down.

A small camera obscura like this can be helpful if you want to sketch realistic pictures. The view you want is shown on the paper with accurate perspective, everything the right relative size and in the right place; in principle, all you have to do is copy it – or even paint directly on the paper.

According to British artist David Hockney, a device like this must have come into use around 1430, because from then on artists in Flanders began to produce pictures that looked more like photographs; in other words, he claims he can detect a change to greater naturalism. Certainly, Leonardo da Vinci gave clear descriptions of a camera obscura in his notebooks in 1490; so the instrument was well established by Tudor times.

However, there is a problem with the simple camera obscura. If the hole is too big, the image is blurred; but if the hole is small enough to make a sharp image, not much light gets through, and so the image is extremely faint. The man who solved this problem was Girolamo Cardano, last seen in the previous chapter with his 'cardan grille' (page 67). In 1550 he described a camera obscura with a lens in the hole. Because the lens focuses the image on the screen, the hole can be much bigger; so it can let in a lot more light and still produce a sharp image. Cardano was a showman; he projected wild scenes of the outdoors into a room full of excited sitters, along with appropriate sound effects. He was certainly a man of wide interests; as well as being a cipher expert, he was Professor of Medicine at Pavia, and in the 1550s he travelled to Britain and cast the horoscope of King Edward VI. Unfortunately he was accused of heresy in 1570 and put in jail, and the story goes that he predicted his own death by means of a horoscope, and then killed himself when his prediction failed to come true.

In 1568 Daniel Barbaro, a Venetian nobleman and architect, published a book in which he described in detail how to set up the camera obscura for drawing. He explained that using a lens makes a much sharper image which can be outlined with a pencil:

> *Close all shutters and doors until no light enters the camera except through the lens, and opposite hold a piece of paper, which you move forward or backward until the scene appears in the sharpest detail. There on the paper you will see the whole view as it really is, with its distances, its colours and shadow and motion, the clouds, the water twinkling, the birds flying. By holding the paper steady you can trace the whole perspective with a pen, shade it and delicately colour it from nature.*

A great popularizer of the period was Giambattista della Porta, who in 1558 produced a book called *Magiae Naturalis* (*Natural Magic*), in which he too explains how to use the camera obscura for drawing. After the publication of this bestseller, David Hockney says, there was a further burst of naturalism from artists – again, images that approached photographic realism. The modern camera is merely an extension of the technology, with the benefit of film or an electronic chip that can capture the image directly.

However, della Porta may well have been more than a popularizer. He and his family were always interested in scientific ideas; he wrote about mathematics and mechanics, and founded a scientific society that met at his home. He may well have had the clever idea of using a mirror to turn the image the right way up, and so make the camera obscura more user-friendly. He explained how to do this in the second edition of his book, and went on to say how it could be used, in a passage that sounds like the first cinema:

> *In a chamber you may see Hunting, Battles of Enemies, and other delusions… opposite to the room where you desire to see this, there must be a large, level space that the sun can shine down upon, where can be placed all manner of trees, forests, rivers, or mountains, as well as animals, and these can be real or artificial, of wood or other material… There can be stags, wild boars, rhinoceroses, elephants, lions and other animals, whatever one wants to be seen; they can slowly creep out of their corners into the space, and then the hunter can appear and stage a hunt.*

These amazing spectacles would certainly have been rather odd if they had all appeared upside-down.

This inventive writer also described how to achieve freezing mixtures by stirring snow with salts, which led some seventy years later to the making of the first ice-cream.

# Pencil power

So from what is now Italy came the use of the camera obscura and a great advance in the technology of art, but there was an important contribution from England too: the discovery that led to the invention of the pencil, which in the long term was probably more important than the perspective devices.

*A lump of pure graphite, or 'wad'.*

The story goes that one stormy night in the 1560s three shepherds, stranded high on the moors of Cumbria, took shelter under a tree that had been uprooted by the wind. In the cold light of dawn they found among the twisted roots a few lumps of a curious black material that they had never seen before. They took some away with them, discovered that it made clear dark marks on things, and used it to mark their sheep. They called it 'wad'.

One drawback to this wad was that it made a terrible mess on their fingers; so the early users wrapped it in cloth or sheepskin, or bound it up with string, to keep their hands clean. The first such writing implements were probably made in Keswick, but the news spread quickly, and one man who heard it was a German naturalist and fossil-hunter called Conrad Gesner who wanted to record his observations on the spot. His charcoal drawings soon smudged, and ink was even worse, but when

*This drawing is from Historiae Animalium book III (1585) by Conrad Gesner, the man who invented the pencil.*

he got hold of some wad he could make clear drawings that survived. He made a stylus by putting a splinter of wad into a wooden handle – and so invented the pencil.

He wrote a book recommending this smudge-free and spill-proof writing implement to fellow fossil-hunters, and pointed out that if you made a mistake you could rub it out and start again. Bread was often used as an eraser, but 200 years later the English chemist Joseph Priestley noted that a particularly effective material for erasing these marks was raw caoutchouc, and suggested that it might be called 'rubber'.

No one knew what wad was made of, but because it was dark and heavy and made marks it came to be called 'black lead' or 'plumbago'. We now call it graphite, after the Greek word *graphein* meaning 'to write'. Chemically it is one form of pure carbon, which also exists as soot, diamonds, and 'buckyballs' or buckminsterfullerene. However, old ideas tend to persist, and we still sometimes talk of 'lead pencils', even though there is no lead in them.

Gesner's stylus or pencil soon found its way into artists' studios, and because the graphite could be sharpened to a fine point, it was perfect for drawing the world in ever-increasing detail. Armed with these new technologies of the pencil and the camera obscura, artists became ever more skilful at making realistic representations of the world. Pictures became much less symbolic and more true to life.

# Mail-order queen

Well aware of this increase in realism, Henry VIII decided to choose his fourth wife from her portrait. Traditionally, dynastic marriages were arranged between (often teenage or even pre-teen) parties who may never even have seen one another before the ceremony – they were pawns in power plays, and personal preference did not have much to do with it. Henry, however, in middle age, felt free to be choosy; he sent an artist to paint pictures of Europe's most eligible young ladies. There were plenty

*For his fourth wife, Henry VIII tried to pick the most beautiful from portraits of the most eligible women in Europe.*

to pick from; indeed, more than a dozen candidates were suggested, including Christina, Duchess of Milan, 'the most beautiful woman in Flanders', who had been married at thirteen and widowed at fourteen, and was said to be a widow and a maid. Her advisers, however, did not approve. Then there were several women in the French court. Henry asked whether he could meet seven or eight of them at Calais, but was told firmly that 'It is not the custom in France to send damsels of that rank and of such noble and princely families to be passed in review as if they were hackneys for sale.' There was the enormous Marie de Guise, as fat as Henry himself had become – his doublets and jackets had to be let out frequently in the late 1530s – but she was already betrothed.

*Above: Henry fell for the miniature portrait, but when he first met Anne he was horrified, and tried in vain to get out of the marriage.*

*Opposite: The pictures he saw of Anne of Cleves were face-on, and did not show that she had a long bony nose and a drooping bosom.*

And then there was Anne of Cleves, who intrigued him, partly, perhaps, because he hoped to get his hands on some alum (see page 198). Anne was German; her brother William had the resonant titles of Duke of Juliers, Gelders, Cleves and Berg, Count of Marchia, Zuphania and of Ravensburg, and Lord in Ravenstein.

The picture of Anne that came back was so stylized that Henry could not tell what she really looked like; his counsellors complained that in the painting it was not possible to see her properly, from which most people assume that they were criticizing not only the ability of the artist but also her rather high neckline – in other words, they were not able to see *enough* of her! So he sent one of the finest artists of the day, his court artist Hans Holbein, to paint her again. Henry was in a hurry and Holbein therefore painted a miniature that was quick and easy to send across Europe. In this miniature Henry thought she was stunningly attractive, and he was smitten. Within weeks he sent a proposal of marriage, and two days after Christmas 1539 she crossed the Channel from Calais with a retinue of 350 people, including thirteen trumpeters.

Henry, having persuaded himself that this was to be his real love, rode romantically to meet her at Rochester Abbey on New Year's Day, and strode unannounced into her chamber. He was confronted by a tall, thin, plain woman, with a very long nose that had not been obvious in the face-front miniature. She wore stiff, old-fashioned German robes, had a tough, determined expression and scarcely a word of English. 'Marvellously astonished and abashed', he forced himself to give her a peck on the cheek, and then left as quickly as he had come.

He tried to wriggle out of the marriage, asking the German envoys for evidence that her prior agreement to marry the son of the Duke of Lorraine had been properly revoked; but they had not brought the necessary papers and he could not put off the wedding. So he married Anne, but disliked her so much he called her 'the Flanders mare', and claimed that he was never able to consummate the marriage. Then he discovered that her previous betrothal had indeed been binding. This gave him an escape route, and his marriage to Anne was dissolved after only seven months, by which time he was already causing a scandal by his frequent meetings with Anne's pretty little maid Catherine Howard.

# Manipulating the image

In spite of Henry's disastrous experience with Anne's portrait, the idea of keeping miniature portraits of loved ones took root. The Tudors also revelled in big formal portraits, which were the pictures that made the most impression; they allowed ordinary people to have some idea of what the King looked like. Portraits of Henry VIII are far more realistic and revealing than even those of his father Henry VII. This revelation was exciting and also dangerous, and before long the possibility of manipulating the image emerged. Henry's daughter Elizabeth drafted laws that prevented anyone from painting her without approval: 'some special cunning paintor might

*The Great Seal of
Elizabeth I.*

be permitted by access to her majesty to take the natural representation of her majesty… he shall have first finished a portraiture thereof, after which finished, her majesty will be content that all other painters or gravors shall and may at her pleasures follow the said pattern.'

Even authorized artists had to work from a handful of approved 'patterns', made by 'pouncing', which meant pricking holes in the outline on a paper pattern, laying it on a prepared board, and scattering carbon dust through the perforations. Then the dots could be joined up. She got the court artist, Nicholas Hilliard, to design a great seal in which she sits in heavenly majesty. The queen had effectively made her image into a logo, and she stamped it everywhere – on coins, in Bibles, on paintings. No one was going to forget who was Queen. And still today the image maketh man and woman – pictures speak more powerfully than words.

# Mapping the country

The greater realism in art was enabled by new technology, and the same thing happened to map-making. Before Tudor times maps had been largely symbolic, but now they became more precise and more useful. A sharp example came in the form of a legal battle between two landowners on the moors of North Yorkshire. All the land around the villages of Old Byland and Murton had once belonged to the local abbey, but Henry VIII had dissolved the monasteries and taken the land. This particular piece he sold to the local families of Wotton and Bellassis. Sixty years later, Sir Edward Wotton, owner of the Old Byland estate, complained that Sir William Bellassis had stolen a stretch of land on the boundary between their estates. The argument became heated and lawyers were called in.

A few years earlier such a dispute might well have been decided by a duel, but science was beginning to make its presence felt, and the authorities sent for a surveyor to settle the argument. The most famous practitioner of the new science of surveying was Christopher Saxton. He listened to the evidence of an old man, John Belwood, who said that when the land was sold by the king, stones were put down to mark a boundary between the two estates. Saxton looked for those stones, and then mapped the area.

The map that Saxton made was presented to a special court at Byland Abbey on 24 August 1598, and it showed conclusively that Wotton had been right: the land had indeed belonged to him. The case was settled, and he got his land back. Four hundred years on, the stone that solved the dispute is still playing its part in modern mapping of the country – as an Ordnance Survey trig point. The Tudors gave us not only modern surveying but also the first really accurate maps.

Twenty years earlier the Queen had decided that her entire kingdom should be mapped, and sent out Christopher Saxton, then only twenty-

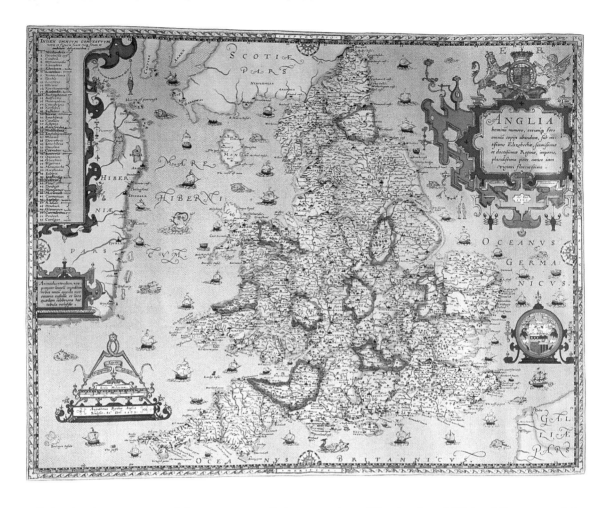

*Christopher Saxton's map of England and Wales.*

eight, to survey all the counties of England and Wales. His maps of the counties of England were published five years later, and were followed by those for Wales. Put together, they became the first complete map of any country in the world.

# Scientific instruments

To help him draw his map for the court, Saxton may well have used a theodolite, which had been invented by Leonard Digges, a remarkable man. Digges had produced in 1556 'A Booke... briefly showing the exact measuring, and speedie reckoning all manner of Land, Squares, Timber, Stone, etc.'. In 1571 he completed another book, published after his death that year by his son Thomas (last seen in Chapter 1 translating Copernicus), in which he describes his invention of the theodolite, an instrument used in surveying for measuring horizontal and vertical angles. Gemma Frisius probably used a similar tool in Germany some years earlier, but Digges was the first to leave us a clear description. Surveyors today use a similar instrument, although now they include telescopes and other refinements.

Digges may also have invented the telescope, for he is known to have experimented with the magnifying effects of combinations of lenses, and he mentions in Chapter 21 of the same book 'the marvellous conclusions that may be performed by glasses concave and convex, or circular and parallel forms', and said he had written a whole book on the 'miraculous effects of perspective glasses'. This book was never published, and we do not know whether he really made a telescope some forty years before Lippershey (see page 168). However, his son Thomas claimed his father used his instrument to read letters and count coins at a distance, and even to spy on people seven miles away. This may have been a bit of an exaggeration, but perhaps it really was a telescope.

So the Tudors produced land maps that were far more accurate than those of their predecessors. However, no Tudor maps were much use at sea. There were fairly accurate charts of the coastline, but sailors did not generally attempt to cross the oceans; they rarely ventured out of sight of land, and they found their way from landmarks on the shore – here a rocky headland, there a church steeple over the trees – until they spotted the harbour they were bound for.

*A replica of Leonard Digges's theodolite.*

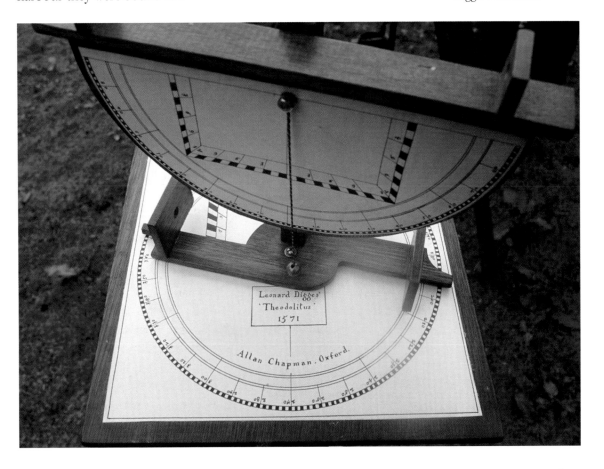

# Braving the oceans

Throughout the known world, there was extensive and complex trade over land: the Silk Route led across Europe, through Constantinople, to the mysterious East, from whence had come rich spices, silks, and precious stones. But the Turks of the Ottoman Empire captured Constantinople in 1453, and almost severed the Spice Route; so the question was, could sailors reach the East by sea? They (or at least the philosophers) knew that the Earth was round – a great globe – and their captains would have been told by men of science that contrary to popular superstition they would not drop off the edge. But was it possible to get to the East by setting off in the opposite direction – west across the open ocean? This was clearly going to need a brave and skilful adventurer, but in principle the rewards would be immense.

*The travels of Christopher Columbus, who never reached the mainland of North America.*

In 1488 a sailor from Genoa, Bartholomew Columbus, came to Henry VII to ask for money to sponsor for his brother's expedition to the west, in search of Japan. His brother Christopher, meanwhile, was trying to extract money from the Spanish and Portuguese authorities. It was going

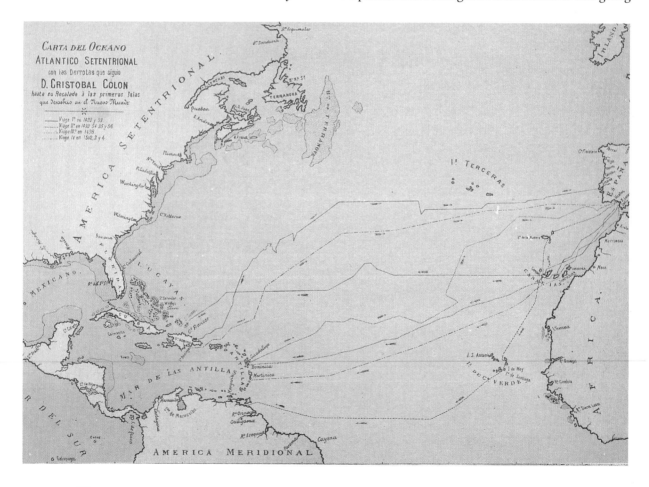

to be a tough trip. Christopher Columbus reckoned he could carry enough food for about a month, and he hoped the distance to the East would not be more than two or three thousand miles, since otherwise his crew would not survive. What he did not know was that the size of the world had been measured accurately by the Greek scientist Eratosthenes 1,700 years earlier, and the distance west from Europe to Japan and China was more like ten thousand miles. Perhaps it was just as well he did not know…

Henry VII could see the point in Bartholomew's expedition, but was not convinced that the goal was realistic, and he hesitated. Eventually, after eight years of looking for sponsorship, Christopher managed to persuade the Spanish government to give him money, and he put together a fleet of three ships. They set off from the Canary Islands on 6 September 1492, and sailed west into the unknown. This must have been scary for him and even more so for his crew of coastal sailors – this time they were out of sight of land for five weeks – but they landed safely in the Bahamas on 12 October. He claimed this new world for Spain, although he thought that the islands were off the coast of China. He cruised around the Caribbean for a few months, and returned early in 1493, after making his sailors swear that they had landed on the Malay peninsula.

*Bronze statue of John Cabot looking pensive on the harbourside in Bristol.*

Even though Christopher Columbus is given the credit for discovering America, he never actually set foot on the North American mainland; the first European to do that – apart from some Viking fishermen – was another sailor from Genoa, Giovanni Caboto, usually known as John Cabot.

## *John Cabot and the* Matthew

Cabot also asked Henry VII for financial support, and this time the king was convinced, and paid up. Cabot set sail in 1496, but ran into terrible storms in the Atlantic, and had to return. On 2 May 1497 he set off again from Bristol – there is a fine bronze statue of him on the harbourside, looking pensive – and headed west and then north-west, across the great open ocean. In these days of high technology, automatic navigation and worldwide communication, it is difficult to imagine how terrifying such a voyage into the unknown must have been. What awful storms might blow up when they had no shelter? What terrible monsters might lurk in the depths,

*A modern replica of John Cabot's ship the* Matthew.

waiting to attack as soon as they were out of sight of land? And if they did reach an unknown shore, what might the hostile natives do?

Cabot's ship was the *Matthew*, a new kind of ship known as a carvel, or later, caravel. At only about 50 tons and 20 metres long, with a crew of eighteen, she was smaller than the cargo ships that were used for trading around Europe – designed for easy manoeuvring in case of danger, for sailing in shallow water, and for anchoring close to beaches so that the crew could go ashore for food and water.

To mark the 500th anniversary of Cabot's great voyage, British yacht designer Colin Mudie planned a replica of the *Matthew,* which was built by shipwrights in Bristol harbour, close to St Mary Redcliffe Church, which according to Elizabeth I was the 'fairest, goodliest, and most famous parish church in England'; we know that Cabot worshipped there. Over the north door inside the church is a model of the replica. The timbers are of oak, apart from the keel. There was no single 15-metre-long straight piece of oak available; so the keel is of opepe, an African hardwood. Below the waterline the planks are of larch; above the waterline they are of Douglas pine, as are the deck and the masts.

The most difficult part of the construction was shaping the planks around the stern; they are 7 cm thick or more, and they had to be bent through 90 degrees in one direction and twisted through 30 degrees in another. This proved to be possible only by steaming the planks for some hours and then quickly clamping them into place. Some of the thicker planks even had to be sawn lengthwise before bending, so that one half could slide past the other. After bending, they were glued together to make a kind of thick plywood. Is this how they built the original *Matthew*? We shall never know, although we do know that the building of the real ship took two years.

The modern *Matthew* is fitted with a diesel engine and a propeller, to get out of trouble if necessary and to manoeuvre in crowded harbours, not to mention radar and GPS, but on her voyage in 1997 she was driven across the Atlantic only by her three sails, which made for a good speed through the water. On 2 May 1997, 500 years to the day after Cabot set sail, she was cheered out of Bristol harbour and down the river Avon by 200,000 people. The weather was poor, and in mid-Atlantic they ran into a Force 10 storm, in which the little ship rolled like a barrel. She did not capsize, but the motion was exceedingly uncomfortable, and at least some of the crew must have wished they had stayed at home. No doubt Cabot's sailors felt much the same. David Alan-Williams, skipper in 1997, managed to reach Newfoundland on exactly the same day as Cabot – 24 June – to be greeted by Queen Elizabeth, the Duke of Edinburgh, and what looked like half the population of Newfoundland. The modern crew had the more difficult navigational task, having to reach the right harbour, but then they also had the technology to know where they were.

On the original voyage, once he was out of sight of land, Cabot had no familiar landmarks to guide him; instead he had to put his faith in science. As long as the Sun shone he could steer west by going away from the rising Sun in the morning, towards the setting Sun in the evening, and keeping the Sun on his left through the day. On clear nights he could keep the pole star on his right. When it was cloudy, however, he had to rely on technology.

He had a simple compass, although it swung wildly to and fro in rough weather, and because they did not understand magnetic deviation, a simple compass bearing could have put them hundreds of miles out by the time they had crossed the Atlantic – though they did not know where they were going anyway, so for this first voyage that was less of a problem.

On a clear day he would have measured his latitude – how far north he was – using either a cross-staff or an astrolabe. The astrolabe is tricky to use on a ship. You have to hold it up, suspended from your thumb, adjust the pointer until you can peer through both peep-holes at the sun, and then read off the angle of elevation on the brass scale. It relies on gravity, and is therefore badly affected by the rolling of the ship. Columbus had both an astrolabe and a quadrant – an earlier instrument based on the same principle – but kept them mainly as status symbols rather than useful instruments, as it's not certain whether he knew how to use them.

*The brass astrolabe relies on gravity, and is tricky to use on-board ship in a rough sea.*

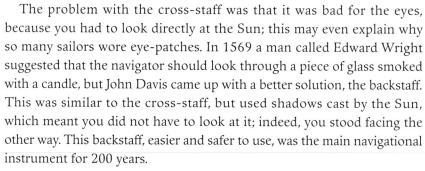

*The cross-staff is a much simpler instrument than the astrolabe for measuring latitude.*

The cross-staff is easier to use than the astrolabe, especially on board ship; you simply hold the end of the shaft by the corner of your eye and slide the cross-bar to and fro until its bottom end lines up with the horizon and its top end with the sun. Then you can read the angle of elevation off the scale on the shaft. At midday, when the Sun is highest in the sky. it will be close to 90 degrees above the horizon for an observer in the tropics, and close to zero degrees for one in the Arctic or Antarctic. Conversely, at night the pole star will be highest in the sky for an observer in the far north, and low in the sky for one near the equator.

The problem with the cross-staff was that it was bad for the eyes, because you had to look directly at the Sun; this may even explain why so many sailors wore eye-patches. In 1569 a man called Edward Wright suggested that the navigator should look through a piece of glass smoked with a candle, but John Davis came up with a better solution, the backstaff. This was similar to the cross-staff, but used shadows cast by the Sun, which meant you did not have to look at it; indeed, you stood facing the other way. This backstaff, easier and safer to use, was the main navigational instrument for 200 years.

## The log

So Cabot could usually find out his latitude, given a clear day. His longitude, however, was much harder. The best method was essentially dead reckoning: to work out how far he had travelled. Measuring distance was tricky. Cabot probably estimated his speed through the water using a log. He literally had a piece of wood tied to the end of a rope, in which were tied knots at regular intervals. The log would be thrown overboard and the man holding the rope would count how many knots slid through his fingers in a fixed amount of time as measured by the sand-glass. Suppose the knots are 101 feet (31 metres) apart, then the speed of the boat through the water is equal to the number of knots that go through your hands in a minute. Speeds of ships and aircraft are still measured in knots – one knot is about 15 per cent faster than 1 mph – and the log lives on as the captain's journal of the voyage; even in *Star Trek* there is a Captain's Log.

By measuring the speed repeatedly through the day and working out the average, Cabot would have had a reasonable idea of how far he had travelled through the water, although the absolute distance might have been affected considerably by ocean currents that he would not have known about.

That first successful voyage across the Atlantic took Cabot fifty-three days. He landed on 24 June 1497, planted a flag in the beach, and claimed this 'new-found land' for the King of England. Wandering a little further inland, he found the remains of a fire, still smouldering, and a carved stick. This was worrying, since the land was clearly inhabited, perhaps by hostile people – and he quickly returned to his ship. For a few weeks he sailed along the coast, and at one point saw shadowy creatures running on the land, but they were so far away he could not tell whether they were people or beasts.

He sailed back to Bristol in the autumn, presented a whalebone to St Mary Redcliffe Church – it still stands there today, in St John's Chapel – and announced his discoveries; Henry VII was delighted and awarded him a pension of £20. Some people believe that the new country was named after Amerigo Vespucci, the Spanish explorer of South America in the early 1500s, but in Bristol the story goes that the king's pension was actually paid to Cabot by the local customs officer, a Welshman called Richard ap-Merrick or ap-Meryke. Ap-Merrick meant son of Merrick, rather like the Icelandic names Ericsson, son of Eric, and Magnusson, son of Magnus, but the Bristolians were confused by the name, and called him Richard Ameryke. Cabot was pleased with his reward, and in thanks said he would name his new-found land after him – America.

*Stars and stripes are visible (top right) in the coat of arms of Richard ap-Merrick or Ameryke in the Lord Mayor's Chapel in Bristol. This may be the origin of the flag of the United States of America.*

Another version of the story is that Richard ap-Merrick was, in addition to being a customs officer, a wealthy businessman or merchant venturer, who sponsored Cabot's voyage, may have helped to fund the building of the *Matthew*, and also organized an audience with the king, asking in return that any new land should be named after him. Cabot's ship *Matthew* may even have been named after ap-Merrick's wife Mattea.

Whether or not either story is true – and what is certain is that ap-Merrick's coat of arms include stars and stripes, as in the American flag – the fact is that Cabot set sail again the following year with a full expedition of five ships and 200 men, planning to consolidate his initial discovery and set up trade with Japan. One storm-damaged ship limped back to Ireland, but the rest were never seen again, and despite various hints and rumours that they had sailed all the way south to Central America, neither Cabot nor any of his sailors ever returned.

# Mapping the world

Having discovered America, the Tudors proceeded to put it on the map – for the first map to show America was drawn in 1507 by a man called Waldsemuller. He had no information about any of the continent apart from the east coast; so his map shows nothing at all to the west. But then in 1520, after another Portuguese sailor, Ferdinand Magellan, had sailed around the world, the west coast appears on the map, although California is shown as an island, and Australia is, as yet, undiscovered. For the first time Europeans knew roughly where the various parts of the world were.

Thus maps, like portraits, were becoming more accurate, but there remained a problem for the sailors who set off across the oceans. When they wanted to head for a specific harbour on the other side, in which direction should they sail? For short distances you could simply lay a ruler on the map and note the compass heading, but that did not work over the Atlantic. Because the Earth is a sphere, and not flat, you cannot lay a ruler across the surface, and even if you have an accurate globe you cannot easily work out the correct course to steer if you want to go from, say, Bristol to Boston.

What they needed was a flat map of the world on which the angles came out right, and that was achieved in 1562 by a Dutchman called Gerardus Mercator. The mathematical process is complicated, but you can make a 'Mercator projection' by putting a lamp in the middle of a transparent globe, and projecting the image on to a cylinder of tracing paper wrapped round the outside, touching along the equator. Imagine that this tracing paper is instant-print film, then the unwrapped print will be a flat map of the world.

Because this was touching along the equator, all the countries in the tropics appear on the map much the same size and shape as they do on the globe. However, countries further from the equator become distorted and

magnified. On Mercator projections Australia and Britain both look bigger than they do on globes, and Greenland looks as big as Africa, whereas in real life it is one-fourteenth the size.

*An 1811 map of the world on Mercator's projection.*

What matters to navigators, however, are the angles, and if you plot a course from Bristol to Greenland on a Mercator projection map, you may be surprised at how small the place is, but at least you will get to the right place. This works because the Mercator projection ensures that all the lines of latitude (around the Earth parallel to the equator) and longitude (through the poles) meet at right angles. Mercator's map was the first to be called an Atlas, after the Greek hero who held the world on his shoulders, and with only slight modifications it is essentially the same map that we use today.

'By these and succeeding Voyages, performed by the Circumnavigators... the Earth was concluded to be truly globous...'

CHRISTOPHER WREN

While Columbus and Cabot were busy discovering America, other sailors were achieving the original goal of sailing to the East. In 1488 Bartholomew Dias sailed from Portugal all the way along the west coast of Africa, and round its southern tip. In July 1497 Vasco da Gama, another Portuguese, set sail from Lisbon with four ships, and rounded this southern cape. The weather was so bad he called it the 'Cape of Storms', but the king later renamed it 'Cape of Good Hope' because of the trading possibilities. Vasco da Gama landed in India on 20 May 1498, and eventually established trade with the country, breaking the Muslim monopoly on spices, and Lisbon became the spice capital of Europe.

Twenty years later Ferdinand Magellan, also Portuguese, fell out with the king, and offered his services to Charles I of Spain, who accepted happily. Magellan had already sailed round the Cape of Good Hope to the Spice Islands in 1511, but the passage round the Cape of Good Hope was now officially controlled by the Portuguese, and so the Spanish were delighted when Magellan proposed to sail west to find another route to the spice trade.

He set sail on 20 September 1519, crossed the Atlantic, and sailed all the way down the coast of South America. Seeing fires along the shore to his left, Magellan named the land there Tierra del Fuego. He spent thirty-eight days feeling his way through what is now called the Strait of Magellan, and on 28 November 1520 sailed into the ocean that he named the Pacific, because it was so calm. By this time one of his ships had been wrecked, and the crew of another had deserted and gone home, and after 100 days' sailing across the vast Pacific Ocean without sight of land the rest of them were desperately short of water and food: they were eating rats and sawdust, and chewing boiled leather. They landed at Guam on 6 March 1521, and after various other adventures completed the voyage all the way round the world, although Magellan was killed in a local battle between islanders. Only one ship survived to return to Spain, but she carried enough spices to pay for the entire expedition.

These heroic voyages finally convinced the landlubbers of Europe that the world was indeed a globe. In Nurnberg Martin Behaim, a map-maker, had begun carving wooden globes to represent the Earth, and marked on them the voyages and discoveries of the navigators. And the idea that the Earth was a sphere, like the Sun and the Moon, helped philosophers to think that it was perhaps just one of many spheres moving through the universe. A hundred and forty years after Magellan's voyage, Christopher Wren said, 'By these and succeeding Voyages, performed by the Circumnavigators... the Earth was concluded to be truly globous... This gave occasion to Copernicus to guess why this Body of Earth... might not move among the Coelestial Bodies.' In other words, these slow and dangerous sea voyages helped to bring about a revolution in astronomy, described in Chapters 1 and 6.

# The first gold rush

Once those pioneering sailors had brought news of strange new lands, all sorts of others saw opportunities of exploiting them in various ways. Martin Frobisher, a Yorkshireman, wanted to find a better route to China than round the southern tip of Africa or South America. He knew that the huge American continent lay in the way of going west, but he wondered whether there might be a short cut, a north-west passage over the shoulder of North America into the China Sea. This looks at least possible on a globe, and he was determined to go and search for it. For about fifteen years he tried to get sponsorship, and eventually, in 1576, he set sail with two tiny ships of about 20 tons each, and a boat that was lost in a storm. His second ship deserted, but he carried on alone. He discovered what he thought was the way through, although it turned out to be merely a deep inlet in what is now called Baffin Island, off the north coast of Canada; it is now called Frobisher Bay after him.

*Sir Martin Frobisher, explorer, who failed to find a north-west passage but did find a mountain of fool's gold.*

More exciting at the time, however, was a cliff on Baffin Island, a cliff of black rock glittering with golden specks. Legend has it that one of Frobisher's men, or more likely the investor Michael Lok, brought back a lump of black stone as a gift for his wife. Disgusted with this ugly gift, she threw it into the fire, at which point it began to glisten like gold.

Gold! The news was electrifying. According to legend, the lumps of black earth Frobisher had brought back were checked and pronounced genuine, but perhaps he lied deliberately, in order to get support for another voyage. He certainly succeeded, for Queen Elizabeth was delighted; here at last was a chance of getting some gold of her own, rather than always having to steal it from the Spanish, who seemed to have control of the gold mines. She gave Frobisher a 200-ton navy ship, established the Cathay company, which gave him the right to sail in every direction but east, and made him high admiral of all lands and waters that he might discover.

In two major expeditions, Frobisher mined 1,400 tons of the black rock, and hauled it 5,000 miles back to a purpose-built factory at Dartford, where hundreds of men toiled for two years, crushing and smelting. But all that glisters is not gold; that effort was entirely wasted, for the glittering specks turned out to be 'fool's gold' – iron pyrites, and almost worthless. All that is left today of that heroic endeavour are a few lumps of stone in the walls in Vicarage Lane, Dartford – the remains of a palace built for Henry VIII – and still, when the sun shines, they glitter like gold. Frobisher went on to sail with Drake (he was knighted for his valour in fighting the Spanish Armada in 1588) and finally with Ralegh. He died in Plymouth after being wounded in a naval battle.

# Spanish gold

There were better ways to make gold. While Martin Frobisher's fantasy turned to dust, his colleague Francis Drake turned to crime. Drake was the first Englishman to sail around the world – and the first person ever to do so and come back alive. He was also a pirate, a slaver and, as far as the Spanish were concerned, a terrorist. In 1577 he was sent off by the Queen for a three-year world tour, and even though he lost his other ships, he ransacked colonies, seized ships, and claimed California for Queen Elizabeth. He came back in 1580 with the *Golden Hind* crammed with gold, silver, jewels and cloves worth, at today's values, something like £25 million. He became a national hero, and the Queen knighted him on board the *Golden Hind*. In 1587 he set fire to the Spanish fleet in Cadiz harbour, and in 1588 he defeated the rebuilt Armada after a week's battle in the English Channel (see page 70). There is a replica of the *Golden Hind* on the south bank of the river in London, and a one-tenth scale model in the museum at Plymouth.

Another sailor, Walter Ralegh (or Raleigh), demonstrated an even cleverer way of making gold. He bet the Queen he could weigh smoke. When she challenged him to prove it, he filled his pipe, weighed it, and then smoked it. Then he weighed what was left. The difference, he said, was the weight of the smoke. Elizabeth thought this was brilliant, and said that although many men had turned gold into smoke, this was the first time anyone had turned smoke into gold. She was fond of Ralegh but when she found out about his affair with one of her maids, Bessy Throckmorton, she sent him to the Tower of London. He managed to get back into favour before the Queen died, but her successor James I liked him even less, and sent him back to the Tower. Ralegh was allowed out for one more expedition in search of a gold mine in South America, but the trip was a disaster, and his head was chopped off in 1618.

# The first colony

Ralegh was rather a poser, and a dandy. He was the man who according to legend got into the Queen's good books when they first met by laying his plush new cloak over a puddle so that she would not wet her feet as she crossed the muddy road; in the town of Raleigh in North Carolina there is a motel called The Velvet Cloak. It was said at one point that he had spent

*The dandy Sir Walter Ralegh, favourite of Elizabeth I and founder of the first English colony in America.*

*A copy of one of John White's watercolours of a native at Roanoake; the caption reads 'The manner of their attire and painting themselves when they goe to their generall huntings, or at theire solemne feasts'.*

6,000 gold pieces on his shoes alone. But he was also a visionary who organized the world's first scientific expedition, and he laid the foundations of the British Empire.

In April 1584 Ralegh sent off an expedition of two small ships to look for a site where he could settle an English colony in America. They found a fine place on Roanoke Island off the coast of what is now North Carolina, where the sailors got on well with the natives (although one of the captains, Arthur Barlowe, was smitten by the chief's wife, and was most disappointed that unlike most of the other women, she covered her chest with 'a long cloke of leather'). Barlowe brought back to England two of the natives from Roanoke, one of whom, Manteo, became a great hit at court, which was good for Ralegh's status. Meanwhile a clever polymath called Thomas Harriot began to compile a phrase-book in the Algonquin language. Harriot taught Manteo English, and in exchange learned about the language and customs of the Algonquin.

In 1585 the Queen knighted her favourite, and in April Sir Walter Ralegh sent off five ships from Plymouth with several hundred people on board, hoping to evade the Spanish and make new lives in the new world, which was named Virginia, in honour of Elizabeth, 'the Virgin Queen'. They met with various adventures on the way, but by the end of August the last ship had sailed for home and 107 colonists were left to fend for themselves.

Along with the sailors, soldiers and farmers, Ralegh had sent the scientist Thomas Harriot and the artist John White. White had started life as a miniaturist, but moved on to use watercolours, which was revolutionary among serious artists. He had his first chance to go abroad with Martin Frobisher in 1577, and his earliest surviving paintings show skin-clad Eskimos paddling kayaks through the ice-floes. These paintings were what persuaded Ralegh to send him to Roanoke.

Thomas Harriot was an extraordinary character. A close contemporary of Francis Bacon (see page 160), he was less of a philosopher and much more of a hands-on scientist and mathematician. He studied at Oxford and was installed by Walter Ralegh as his mathematical tutor at Durham House on the Strand; Harriot's job was to teach Ralegh's sailors navigational science. Ralegh sent him as surveyor to Roanoke in 1585, and he surveyed the area around the new colony – presumably with a theodolite; he produced a remarkably accurate map, and in 1588 published *A Brief and True Report of the new-found Land of Virginia*. This little book described the native customs of farming, foods, animals and plants. Meanwhile, White sketched all these things in pencil – an early devotee of the new technology from the Lake District – and then painted them in beautiful coloured detail. These were the first pictures seen outside America of flamingos, iguanas and other weird and wonderful things. Most important of all, though, he sketched and painted the people.

Igwano. Some of thes are 3. fote in length. and lyne on land.

*John White's paintings were probably the first clear images of exotic people and animals anyone in England had seen.*

Harriot may well have been using a telescope; we know the Tudors were experimenting with lenses and mirrors, and Harriot says he took to Roanoke a 'perspective glass whereby was showed many strange sights'. Some historians of science believe that Harriot had followed Digges (page 87) and made a telescope, twenty-five years before it was invented by Lippershey and developed by Galileo (page 168). He observed comets, the moons of Jupiter, and the lunar surface, which he drew in 1609. He seems to have been the first person ever to observe sunspots, and from their movement he worked out the rate of rotation of the Sun. He corresponded with Kepler between 1606 and 1609; built himself an observatory, and had the largest astronomical instrument in England – a 12-foot cross-staff.

Harriot also made notes on a huge range of practical things: the best size for a ship's mast, the maximum supportable population of the Earth, the gaseous yield of a burning candle, clever new methods of navigation, the densities of a range of materials, including gold, silver, antimony and lobster shells, and a study of the velocity of water flow, which led to a new plumbing system for Petworth House, home of the Earl of Northumberland, who became his patron.

He was a smoking enthusiast, and tried to persuade his reluctant contemporaries that tobacco had useful medicinal properties. He got his comeuppance, though, for he died from a terrible black tumour on his nose; he was probably the first man in England to die from smoking.

In 1587 Manteo returned to Roanoke with a new batch of colonists, but now he was Lord Manteo, the chosen representative of the Queen of England, which was to say the least a controversial appointment. In 1590 a further group of English visitors, led by the artist John White, who had been made Governor, found the colony abandoned. No one was able to find out whether the colonists were killed, had starved to death, or simply moved on somewhere. Mysteriously, the name of Manteo's village, Croatoan, was carved on a tree.

Ralegh's colony survived for only about five years, but it provided the old world with its first look at the new; moreover, the colony's existence may even have been the reason why Americans speak English to this day, rather than Spanish, or even Portuguese.

One of the lasting influences of the Elizabethans on world affairs was the founding of the British East India Company, which was given a royal charter by the Queen on 31 December 1600. 'The Governer and Company of Merchants of london trading into the East Indies' were granted the exclusive right to trade in India, Africa, and America, and in 1610 and 1611 they set up trading posts in Madras and Bombay. The growing influence of the company as trade steadily increased was one of the major causes of the British domination of India. Robert Clive, of the East India Company, won decisive battles at Arcot and Trichinopoly in India against the French in the middle of the eighteenth century, and in 1773 the British Government appointed a Governor General of India. In other words, India became part of the British empire as a direct result of Tudor trade.

So what the Tudors did for us in this sphere of artistry and exploration was to bring about new ways of seeing the world, and then to record those images with a new precision. They made the first accurate maps and developed scientific instruments to aid surveying and navigation on land and sea. Through hazardous ocean voyages, important trade routes were opened up and new lands discovered; the first attempts to settle in these lands laid the foundations of the British Empire.

Being the Fourth Chapter

# THE GOOD LIFE

'11 entire beef carcasses, 6 sheep, 17 pigs,
500 chickens, 15 swans, 6 cranes, 380 pigeons,
600 larks, 70 geese, 4 peacocks, 3,000 pears,
1,300 apples, and 3,000 loaves of bread'

THE INGREDIENTS FOR A SINGLE FEAST HOSTED BY HENRY VIII

*Previous spread: a detail from a tapestry at Hardwick Hall, with the owner Elizabeth Shrewsbury's initials.*

While great revolutions in science and religion were sweeping through Europe and transforming the way that Tudor people saw the world, other changes were happening on a more everyday level, affecting all aspects of daily life. Many of these changes, both home-grown and imported from abroad, have echoed down the years, building up an appreciation of the quality of life. No matter how many centuries pass, people still like to live in pleasant homes, eat and drink well, and wear fashionable clothes; they like to play games and be entertained. Certainly, on a basic level our needs don't change, and few things are as basic as a traditional Tudor lavatory – though even here, there were signs of improvements to come.

# The jakes progress

We now happily take sewage systems for granted, disposing of our waste at the flush of a cistern. Our Tudor ancestors had no such luxury. Like generations before them, they used a privy: usually a little wooden hut with a wooden seat inside. There would be a hole in the seat, below which was a pit to contain the sewage. Toilet paper had not been invented, and indeed paper was a rare commodity anyway; so most people probably used leaves or handfuls of moss. When the hole filled up, they would dig another a few yards away, move the hut over the new hole, and put a spadeful of earth on the old one. Out in the country, people would put their privies at the bottom of the garden or in the back yard; in the more crowded towns, there could be common privies serving a whole street.

*The garderobe at Acton Court – an ensuite lavatory which was apparently widened for its royal visitor, Henry VIII.*

No one would want to go out to the privy at night, when it was dark and perhaps raining, so people kept chamber pots under their beds and then in the morning emptied the contents out of the window, with a warning shout of 'Gardy loo!' – from the French *Regardez l'eau!* meaning 'Mind the water!' (And our slang word 'loo' is a probably a memory of this, too.) The streets in towns would have run with raw sewage, along with manure from animals, everything churned by passing traffic.

Many privies were 'one-and-a-half-seaters' – that is, they had one ordinary seat for an adult and a lower one for a small child, so that Mum could go with a toddler. Some, however, were much bigger; there is even a six-seater at Chilthorne Domer Manor in Somerset; presumably the whole family used to go together. This is rather odd, since the word privy comes from the Latin *privatus*, meaning private, and it can't have been very private if you went with five other people!

There was room for even more in Hampton Court's 'Great House of Easement', for the relief of Henry VIII's courtiers. This was a two-floor public lavatory with a wall down the centre. On each floor, against each side of the wall, was an oak bench with a row of seven holes. Perhaps women used one side of the wall and men the other. In any case it must have been cosy, since the holes in the bench were only about two feet apart, and there appear to have been no partitions between the sitters. All the sewage from this Great House of Easement trickled down angled chutes into a brick vault below but, amazingly, there seems to have been no flushing system to get it out into the river. At regular intervals the gongfermors or gongscourers – men armed with shovels – were summoned to clear it out. Attacking the head-high mound would not have been a pleasant task, especially in high summer.

Kings and queens would certainly not have had to share facilities, much less actually venture out to privies. Henry VIII used a handsome 'close stool', which was a commode containing a pewter pot. It was sumptuously upholstered in black velvet, and decorated with fringes, ribbons, and 2,000 gold nails. One of the senior courtiers had the privileged rank of 'Groom of the Stool', and this gentleman left us a charming report after the King had taken laxatives one night in 1539; he was summoned at 2.00 am, 'when His Grace rose to go upon his stool, which, with the working of the pills and enema… had a very fair siege'. And no common moss or leaves for the King: he would have had the use of linen or flannel, perhaps lambswool.

Even when he was travelling, his hosts did their best for him. At Iron Acton near Bristol, the wonderful house of Acton Court was built around 1534 by Sir Nicholas Poyntz apparently just for a royal visit; Henry took Anne Boleyn there for a weekend. The house has recently been restored, and one of the major discoveries was a garderobe – an *ensuite* lavatory in the shape of a cupboard (which appears to have been widened for the royal visitor), containing an oak seat with a hole over a 22-foot drop down a stone shaft into the moat.

Elizabeth would also have had her own richly upholstered close stool, but in one respect she may have been luckier than her father, for her godson Sir John Harington designed her a self-contained water-closet, complete with cistern and flush. He was immensely proud of the concept, and wrote a learned and amusing book about it called *The Metamorphosis of Ajax*. This was a pun – the common Elizabethan word for a lavatory of any sort was a 'jakes', and his book was about building a new lavatory; or metamorphosis of a jakes.

'His Grace rose to go upon his stool, which, with the working of the pills and enema… had a very fair siege'

REPORT BY HENRY VIII's GROOM OF THE STOOL

*Sir John Harington, court wit and designer of the self-contained flushing 'loo'.*

Some people suggest that Harington actually invented the water-closet, but this is not true; the Romans had had flushing closets 1,500 years earlier, and even they were not the first. However, his device seems to have been an original invention, and it may have been the first design for a self-contained water-closet, at least in Britain. According to the English biographer Lytton Strachey, Harington's design was a hit with the Queen – though she already had a soft spot for him: 'She liked the foolish fellow. She had known him since he was a child; he was her godson – almost, indeed, a family connection, for his father's first wife had been a natural daughter of her own indefatigable sire.'

Through his wife, Harington had inherited his fine Italian house at Kelston in Somerset, where one day Elizabeth honoured him with a visit:

'He had felt himself obliged to rebuild half the house to lodge his great guest fittingly; but he cared little for that – he wrote a rhyming epigram about it all, which amused the ladies of the bedchamber. He wrote, he found, with extraordinary ease and pleasure; the words came positively running off the end of his pen; and so – to amuse the ladies again, or to tease them – he translated the twenty-eighth book of Ariosto's *Orlando Furioso*, in which the far from decorous history of the fair Fiametta is told. The Queen soon got wind of this. She read the manuscript and sent for the poet. She was shocked, she said, by this attempt to demoralize her household; and she banished the offender from court until – could there be a more proper punishment? – he should have completed the translation of the whole poem...'

The ladies certainly enjoyed his translations of naughty foreign verses. Another example is his version of a poem about smells – echoes of the jakes again – from the Latin original by Thomas More (here most of the words have been modernized):

> *If leeks you leeke, but do their smell disleeke,*
> *Eat onions, and you shall not smell the leek:*
> *If you of onions would the scent expel,*
> *Eat garlic; that shall drown the onion's smell;*
> *But against garlic's savour, if you smart,*
> *I know but one receipt. What's that? A fart.*

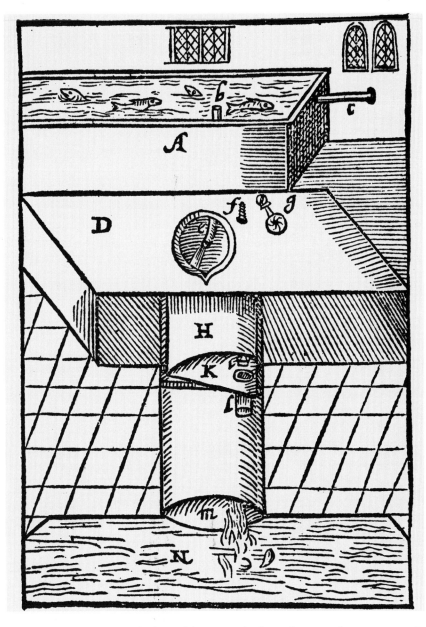

*Diagram from Haringtonʼs book* The Metamorphosis of Ajax *shows how his lavatory worked: note the goldfish in the cistern.*

'...how unsaverie places may be made sweet, noysome places be made wholesome, filthie places made cleanly.'

SIR JOHN HARINGTON

Harington's *Ajax* concludes with a practical section, on 'how unsaverie places may be made sweet, noysome places be made wholesome, filthie places made cleanly'. Such references clearly show that Tudor people did not embrace dirt and squalor. They probably did their best to keep themselves, their clothes and their environment as clean as possible with very limited resources.

To keep clean, you need water. Piped water was rare in Tudor houses, even the grand ones. Most people had to fetch their water from a well or a pump, and dirty water was often emptied straight on to the ground. It was not until the middle of the nineteenth century that drinking water was completely separated from sewage, with water piped into houses and main sewers to remove the effluent. Contamination from sewage was the main reason why so many babies died throughout history – from diarrhoea and similar infections. There were no official records kept in Tudor times, but it has been estimated that one in five newborn babies would not survive to see their first birthday. In the second half of the sixteenth century, a quarter of those children who did survive that perilous first year would die before they reached ten years of age. In the crowded slums of London, mortality rates were even higher.

On the other hand, those who did survive their early years often lived to a ripe old age, no doubt having acquired a good deal of immunity when they were exposed to so much disease and dirt. Also, early in the sixteenth century there seems to have been a fashion for healthiness. Henry VII somehow became persuaded that physical fitness was desirable. Being a cautious, thrifty monarch who would not willingly spend resources on warfare, he began to arrange alternative physical activity in the form of organized sport.

# The sporty good life

Henry built a sports complex at Richmond – the first ever recreational buildings at an English royal palace – which included billiard tables, bowling alleys, archery butts and tennis courts. They also had viewing areas for spectators; presumably the ladies of the court were not expected to take part in such strenuous activity. The king apparently believed that being good at sporting activity demonstrated physical grace and fitness, and also provided the opportunity for competitive displays.

Britain today is a nation of sports fans. Some of them actually enjoy playing games, but many more are only too happy to sit in front of the TV and watch others exhaust themselves on every sort of arena from the racecourse to the football field. This great enthusiasm for sport began with the Tudors. Henry Tudor approved of sport, but his son Henry VIII was a fanatic. In his youth he had spent the afternoons in martial arts – jousting at the rings, fighting with blunt swords on foot and on horseback, practising with the longbow, running races, leaping ditches and wrestling. In the last few years of his father's life he had been somewhat held back by his father's protective attitude, but when he became king in 1509 at the age of nearly

eighteen he launched himself into a wild round of sporting activity, and was the equal of anyone at jousting and throwing the spear (see page 53).

He built indoor sports centres at Hampton Court, Greenwich, and especially Whitehall, where there were four tennis courts, two bowling alleys and a cockpit, as well as a large

tilt-yard for jousting, and a park stocked for hunting. As a result of all this, the young men Henry gathered around him at court became fit, healthy and ready for battle – which was part of the bellicose king's plan. He insisted that they should practise archery every day, and in 1512 introduced a law forbidding ordinary people from playing such idle games as cards, dice, bowls, and skittles – which was presumably for both moral and practical reasons. There is a story that the game of darts was invented or at least popularized by the Tudors in order to allow the archers to keep practising even in the dark winter months. Certainly Anne Boleyn presented Henry with a beautiful set of darts in 1530, while she was still trying to win his heart. She scored a bull's eye, and the game of darts has been popular ever since.

One of the games that flourished and developed at this time was tennis – now one of the major stars of the sporting firmament. Royal or 'real' tennis was played indoors, in a stone court with a sloping roof on one side. The ball was made of sheepskin stuffed with sawdust or wool – rubber balls would not be invented for another 300 years. To begin with, the players hit the ball with their hands, as we do today in fives and volleyball; the game originated among French monks in the twelfth century, and was called *jeu de paume* or 'palm game'.

*The rackets for real tennis probably started as kitchen sieves, which is why they were strung diagonally.*

*With its magnificent four-poster bed and tapestries, the Blue Bedroom at Hardwick Hall illustrates how the Countess of Shrewsbury made her new home the lap of luxury.*

They gradually developed various techniques to hit the ball harder. One theory is that they made a mesh of string to loop over their fingers, and this developed into the tennis racket. Another, more probable, idea is that they used a sieve to serve the ball – sieves of string tied in a mesh over a circular wooden frame were common kitchen utensils – and then put a handle on it to make a racket, more or less as we have it today. Tudor rackets were eventually made of yew wood and loosely strung with catgut; there was none of the power of today's equipment, and no question of the 100 mph serve. The curious scoring system of 15–love, 40–15, deuce and so on was all part of the Tudor game. One school of thought reckons that this system was based on the quarters of a clock: 15, 30, 45 (later reduced to 40), with 60 as the final point. A nice (but unsubstantiated) theory is that 'love' got its name because a written 'nought' looks like an egg; the French for egg is *l'oeuf* – and given the traditional English ineptness with foreign words…

An additional attraction of the game for spectators was gambling, a favourite Tudor pastime; spectators often laid large bets not just on the result of a match but on every point. Henry would make bets on everything, but the thrill of the gambling sometimes took over, and he would wager ridiculous amounts of money, especially on dice. As a result all sorts of shady characters appeared at court, apparently boundlessly wealthy and keen to take on the king; they were really professional gamblers, only too skilful at losing a little and then winning a lot. For some years Henry was too naïve to see what was going on.

One of Henry's queens was a great gambler too. The story goes that as Anne Boleyn was led off to the Tower, she complained bitterly that she had not had time to collect her winnings on a tennis match; she won her bet, but she lost her head.

The enthusiasm for fitness and self-improvement was reflected in some of the books published at the time. There were self-help books on etiquette, mental improvement and, in 1542, *A Compenyous Regyment, or a Dyetary of Health,* which recommended regular healthy exercise. Many of the grand houses that were built at this time had enormous long galleries that were partly places to hang the family portraits and partly indoor walks, so that you could still get your exercise even in bad weather. Hardwick Hall in Derbyshire, built by the Countess of Shrewsbury in the 1590s, has the longest gallery in England: 170 feet (56 metres).

*Hardwick Hall was built to be 'more glass than wall'.*

The Countess, familiarly called 'Bess of Hardwick', was the richest woman in England after Queen Elizabeth. She married four times, and became immensely powerful; her sons began the Cavendish dynasty, and the Dukes of Devonshire are direct descendants. She built Hardwick Hall in her seventies as a proud statement of her standing, demanding that it should have 'more glass than wall', and it is now one of the finest examples of Elizabethan architecture. The level of luxury in these grand houses represented an enormous advance from the dark and draughty medieval structures. After centuries of internal strife, peace at home meant that great houses could be more like stately homes to be enjoyed and less like fortresses to be defended.

Bess certainly developed a taste for the good life; she had padded furniture, a fireplace in every room and floor-to-ceiling glass windows. Many people at this time were beginning to use wallpaper. This novel idea of covering walls with paper was one factor in kick-starting the first full-size paper mill in England in 1595. The Master's Lodge of Christ's College Cambridge was found in 1911 to contain a fragment of wallpaper dating from 1509; it had an H and a picture of a bird and was probably made by a man called Hugo Goes. The fashion took most of the sixteenth century to develop, but eventually it really caught on, and Britain became a major producer of wallpaper in the eighteenth century.

Bess even built a special room on the roof, to which she would lead her guests for their dessert. They had to teeter in their elaborate costumes over a rough wooden walkway, but were rewarded when they arrived with superb views of the estate. They were rewarded also with wonderful food.

# The gastronomic good life

One problem for athletes is that they have to eat a great deal to maintain the energy and the capacity for exercise, but when they give up the sport they are liable to carry on eating, and get fat. In the late 1530s Henry swelled like a balloon – his clothes all had to be let out – and this must have been mainly because of his prodigious appetite. Poor people lived on a meagre diet of dark bread, broth and perhaps cheese, but the rich really let themselves go.

When the Tudor courtiers ate, they did so with enthusiasm. Some said that English stomachs needed more food than others, and the English had a habit of shameless belching at the table, although good manners and etiquette were important; messing about with food was strictly disallowed, and children had to be taught not to chew the bones. In 1517 banqueting laws were passed which limited to seven the number of courses a churchman or noble could serve his hundreds of guests, but since each course might include a dozen dishes, this was not a serious restraint. Luckily the king was in any case exempt from such restrictions, and often he entertained a thousand or more people. In a single day's feasting he and his guests once ate 11 entire beef carcasses, 6 sheep, 17 pigs, 500 chickens, 15 swans, 6 cranes, 380 pigeons, 600 larks, 70 geese and 4 peacocks, not to mention 3,000 pears, 1,300 apples and 3,000 loaves of bread.

These gargantuan meals were served with incredible pomp – for the king, at least – and with immense ritual and tradition. Each sort of meat had to have the correct sauce, and the dishes had to come in the right order. Between courses of pastries stuffed with venison, capon pasties, beef marrow fritters, eels in spicy sauce, sturgeon's liver, calf's liver, lampreys with hot sauce and whale meat came elaborate sculptures of sugar and wax depicting, say, biblical scenes, placed in the middle of the table. Meanwhile chaos reigned in the hall: guests' servants would be running around carrying messages, while the host's servants would be rushing in with more food and out with empty dishes; the posh people's tasters tried all the food to make sure it wasn't poisoned, dogs snarled and fought over scraps and bones, and the evening's entertainers – jugglers, musicians, or acrobats – had to compete with everything else to make themselves noticed.

Diners would eat from trenchers, which in the Middle Ages had been slabs of bread, but by Tudor times were dishes of wood – or of earthenware or metal for grand people. They filled their trenchers from the great plates of food in the middle of the table. The grandest people had a great variety

of dishes at the same time – starters, main courses and puddings were all served together – and would have to share with only one or two other people. Down at the other end of the hall several diners might be sharing from just a few plates, and therefore eating rather simpler food.

Sugar, sweet things, and spices were fairly new – they were all recent imports, products of the growth in foreign trade (see page 88) – and therefore exceedingly fashionable. Most meat courses were sweetened with pears or sugar; chicken might be roasted but then glazed with a sticky sauce flavoured with pears, oranges and cloves, and decorated with nasturtiums. Pasties were stuffed with trout, flavoured with ginger, cinnamon, sugar and dates. Jellied cubes of milk and rosewater were covered with gold leaf. Even the wines were often sweetened with honey and spices, and sugared wine was thought to be an aphrodisiac.

Sugar brought about a revolution in the Tudor kitchen; they liked it so much they invented a separate course for it, and called it the banquet. We now call it dessert after the past participle of the French verb *desservir* – to clear the table. After the main course the grand people would go into a separate room to eat it while the servants cleared the main table.

Henry VIII enjoyed feasts anywhere, and in 1518 he was entertained to an extraordinary banquet in Venice – aboard ship. Three hundred courtiers were rowed out to the flagship and took their places at the long tables on the

*The Tudors loved their food, and were particularly enthusiastic about the newly-imported luxury ingredient, sugar.*

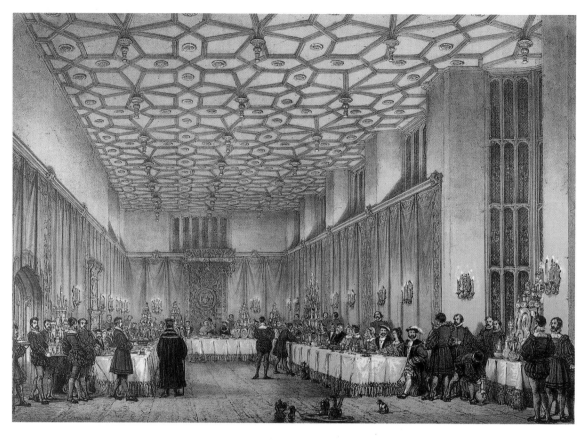

*Cardinal Wolsey hosts a modest little feast at Hampton Court in the 1520s.*

deck, while the king and the Venetian Ambassador sat on a raised platform looking down on the others. The masts were festooned with silks and tapestries, and the tables were groaning with cakes, sweets and wine. As they ploughed their way through the meal, they were entertained by sailors swinging acrobatically through the rigging overhead. Henry was enormously impressed, and lavish with his praise.

Along with sugar and spices, exotic new foods began to appear on the table. The potato, first cultivated in South America thousands of years ago, was brought back by Spanish sailors from the West Indies around 1570, and gradually spread across Europe. One story is that potatoes were introduced to England by Sir Thomas Harriot in 1586, but curiously the myths about potatoes seem to have spread faster than the plant; for many decades they were called 'Virginia potatoes' and were supposed to have been introduced to Ireland by Sir Walter Ralegh, after his trip to Virginia. In fact he never went to Virginia, and potatoes were not cultivated there in the sixteenth century; so that story is entirely fictitious. For a long time potatoes were regarded with suspicion, possibly because they resemble plants of the nightshade family, and considered fit only for animal fodder. Even after the Royal Society recommended them as useful food in 1662, another hundred years went by before they were widely eaten.

Tomatoes also originated in South America, and were brought back to Europe by Spanish sailors, rather earlier than potatoes. By the 1540s they were being eaten in Italy with oil, salt and pepper, and by the end of the sixteenth century they were being eaten with caution by people in England.

Tobacco caught on rather more quickly than potatoes and tomatoes; it was introduced from America by Sir John Hawkins in the 1560s (and, in the process, secured the economic viability of the colony of Virginia). Walter Ralegh, always in the vanguard of fashion, helped to popularize smoking at court, using long-stemmed clay pipes, and soon both men and women of all classes were puffing away. In 1573 it was reported: 'The taking-in of the smoke of the Indian herbe called Tabaco, by an instrument formed like a litle ladell, whereby it passeth from the mouth into the hed & stomach, is gretlie taken-up and used in England.' Ironically, in view of its status centuries later as a major health hazard, tobacco was initially valued as a medicinal plant, positively beneficial to the system. (Though even in the middle of the twentieth century at least one major brand of cigarettes was advertised with the slogan: 'For your throat's sake, smoke!')

A smoke and a pint long went together: while the rich had rare and precious wines, most people drank ale. In fact there was very little choice of beverage in Tudor England: drinking water could make you ill, and milk was too important a commodity for regular drinking. In fact, ale was drunk by just about everybody, of all ages, and had been the traditional drink in England for centuries. Ale was usually brewed by the lady of the house, using methods still familiar today – though of course now much refined with hi-tech machinery. She would use malted barley (barley that had germinated and then been baked), water and yeast. It might be flavoured with thyme or rosemary, or nettles or spruce needles. Fortuitously, because the boiling process killed bacteria, it was much less dangerous than water. A man could well drink a gallon or so in the day, depending on the strength of the ale – that is, how many times the brewer would reuse her malted barley.

Ale was originally distinguished from beer, which was brewed in a similar way but with the addition of hops. Here too the malted barley could be used again and again, producing progressively weaker beer; the weakest was known as 'small beer' – a term still used today for anything unimportant. In 1516, the German Purity Law (allegedly the world's first food regulation) decreed that beer could be made only from water, malted barley, yeast and hops. The use of hops would have a long-lasting effect on English brews: the natives have been drinking hopped beer ever since. Hops

'The taking-in of the smoke of the Indian herbe called Tabaco, by an instrument formed like a litle ladell, whereby it passeth from the mouth into the hed & stomach, is gretlie taken-up and used in England.'
1573 REPORT

not only give beer a bitter taste but, more importantly, help keep it clear and act as a preservative. At first people resisted the idea of adding hops, but when the brewers realized that beer with hops could be kept for longer, they jumped on the bandwagon, and the old-style ale began to die out. (The old distinction between ale and beer died out too, and today the terms 'ale' and 'beer' are interchangeable.)

# Entertainment

Apart from eating and drinking (and smoking), the Tudors were enthusiastic merrymakers: they liked to entertain and to be entertained. Henry VIII loved not only sport and food, but also music. He played several instruments, including bagpipes, recorders and flutes, and the virginals. His Chapel Royal included dozens of musicians, many of whom used to accompany him wherever he went. Church music was generally written to be sung unaccompanied, but Henry liked to have trombones to accompany the plainsong of the choir.

Henry composed various ballads including, according to legend, 'Greensleeves'. His daughter Elizabeth was also a skilled musician, and especially fond of playing the virginals. Most rich girls were taught to play instruments when they were young, and would later entertain themselves and their families, perhaps after their evening meal. Poor people would hear the music in church, and also gather round to listen to travelling musicians or minstrels. Anyone could afford to make a simple recorder or penny whistle; so home-made music was probably fairly widespread.

Some of the Tudor instruments have survived to this day, including the trumpet, recorder, flute, bagpipes, bells and cornet. Others disappeared, including the virginals, the crumhorn, the shawm and the rouschfife, which were rather like oboes, and the dulcimer, which was a sort of cross between a trapezoidal guitar and a primitive piano; the player hit the strings with hammers, giving a powerful sound, so that the dulcimer could be played outdoors.

The Tudor period was remarkable for its array of musical talent, and one of its greatest musicians was Thomas Tallis, who was born in 1505, called to the Chapel Royal by Henry VIII, and remained in royal service for Edward, Mary and Elizabeth. A master of counterpoint, he has been called the 'father of English church music', writing anthems, chants and hymn tunes that are still in use today, such as his setting of the Canticles in D Minor.

Another member of the Chapel Royal was John Heywood, born near St Albans in 1497. He married Sir Thomas More's niece, Elizabeth Rastell, and as a result of his great wit and skill became a favourite of both Henry VIII and Mary, although because he was a firm Catholic he was less popular with the Protestant Edward VI. As well as being a musician,

Heywood was a clever writer of epigrams and proverbs, of which a great many survive to this day:

> All's well that ends well.
> A man may well bring a horse to the water, but he cannot make him drink.
> A penny for your thoughts.
> Beggars shouldn't be choosers.
> Better late than never.
> Butter would not melt in her mouth.
> It's an ill wind that blows no good.
> Look before you leap.
> Many hands make light work.
> Out of the frying pan into the fire.
> Rome was not built in a day.
> Two heads are better than one.
> When the iron is hot, strike.
> When the sun shineth, make hay.

Heywood also began writing plays, and in particular was probably the first playwright to turn the abstract characters of the medieval morality plays into real people. His farces were immensely popular, and may well have contributed to the growing interest in plays as a form of entertainment. Perhaps the Protestant reformation of the Church, which stripped out much colour and ritual – and indeed drama – from people's lives, increased the

*Many Tudors, including Henry VIII and his daughter Elizabeth, enjoyed playing the virginals, which were like a primitive harpsichord.*

*William Shakespeare, literary genius and best-known playwright of the Tudor period.*

appetite for theatrical performances. In a world without cinema or television, theatres would give people a new shared experience.

The first proper theatres in London were built in 1576 at Finsbury Fields in Shoreditch and at Newington Butts in Southwark (where Michael Faraday would be born 215 years later). London was by now the biggest city in Europe, which meant that there were enough people to fill theatres, and so make money for impresarios and actors. These first theatres turned out to be successful, and four more were built before the turn of the century, including the Rose and the Globe, not far apart on the south bank of the Thames.

The timing could hardly have been better for a young man called William Shakespeare, who was born in Stratford-on-Avon in 1564 and moved to London in his late twenties. He seems to have become an actor around 1591, and began writing at about the same time, although he probably wrote more in the following two years, when the theatres were closed because of the plague. He became one of the Lord Chamberlain's company of players, who later became 'The King's Men'. They built the Globe, and he became a partner; this was where all his greatest plays were produced, although in 1613 the Globe burned to the ground when a cannon was let off – as a special effect – and set fire to the roof.

Shakespeare's command of English was extraordinary, and he gave us a wealth of phrases and ideas that will never fade. He crowned the flowering of dramatic genius in the Tudor period, which saw such other luminaries as Christopher Marlowe, Thomas Kyd and Ben Jonson. One of the most significant things the Tudors did for us was to leave an immortal body of drama and poetry.

# Keeping up appearances

If playwrights and theatres could put on a show, so could the people themselves. For Tudors in a position to afford it, personal appearance was very important. This could be linked to hygiene – as we have seen, people would do their best to keep themselves clean, in difficult circumstances.

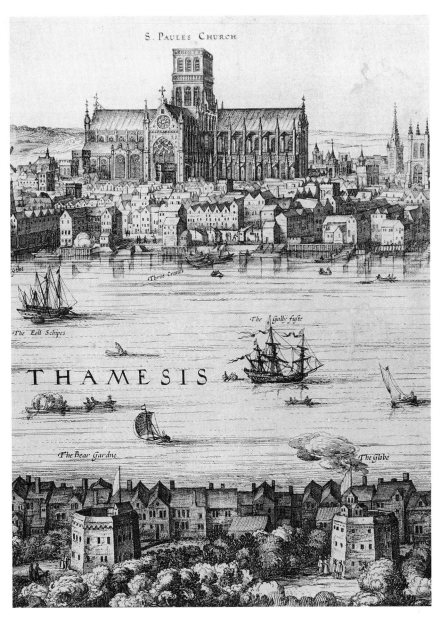

S. PAULES CHURCH

THAMESIS

The Eell Schipes

The Gally fuste

The Bear Gardne

The Globe

'All the world's a stage, And all the men and women, merely Players...'
WILLIAM SHAKESPEARE

They tried to look after their teeth, too: the toothbrush was allegedly invented in China around 1500, and reached Europe just after the Tudor period; in 1649 Sir Ralph Verney was asked to bring toothbrushes back from Paris. Tudor children were taught to clean their teeth with toothpicks, and to rub them with a linen cloth to whiten them; in order to keep their breath sweet they were told to sleep with their mouths open. However, in a move that would strike us now as bizarre, Tudor ladies would often go to the other extreme, and deliberately blacken their teeth. This was supposed to show that they were wealthy enough to have lots of sugar, and it was well known that sugar rots the teeth.

*The Globe theatre (bottom right) with the Bear Garden (bottom left) and the old St Paul's across the river.*

In general, however, Elizabethan ladies took immense trouble over their personal appearance. In the middle of the sixteenth century mirrors were rare luxuries, following the Puritan influence which scorned self-adornment, but by the end of the century they had become necessities; a pocket mirror of glass or polished steel became an essential item of every courtier's effects. Queen Elizabeth herself took plenty of time being made up, and wore a massive red wig; she loved dancing, and used her beauty and her unmarried state as weapons in the power-play at court, juggling her suitors from home and abroad, and glorying in her image as the half-divine Virgin Queen.

*Queen Elizabeth used her appearance as a powerful weapon in all aspects of her reign, bolstering her image as the half-divine Virgin Queen.*

The most important single factor in personal appearance was to have pale skin, in order to demonstrate that you did not need to work and expose yourself to the sun. So both men and women used emollients of various kinds – lemon juice and rosewater to bleach the skin, mercury to make it smoother – and whitening make-up made from alabaster or powdered white lead, sometimes stuck on with egg-white. This face-powder had the additional advantage of covering any pock-marks left by smallpox, which most people caught at some stage in their lives. They did not realize that lead is poisonous, especially when you are exposed to it for long periods of time, and some people even died of lead poisoning just from their face-powder.

In addition they used blusher or rouge made from the same sort of face-powder with added dye, usually red ochre. For a lipstick, they ground up plaster of Paris and mixed it into a paste with a colouring material, rolled the mixture into a crayon, and dried it in the sun.

The Tudors were increasingly enthusiastic about their clothes. Fashions spread into England from the Continent – there is clear evidence in Elizabeth's wardrobe accounts – and courtiers took to displaying their wealth on their backs, Walter Ralegh being one of the most extravagant (see page 98).

Next to their skin, women wore a chemise or smock, which was rather like a long-sleeved nightdress, usually made of silk for the rich or linen for the poor, and sewn together from simple square pieces without any gathering at the neck or wrists. The chemise was designed to protect the rest of the clothing from sweat, and since Tudors rarely had a chance to wash, it must have become quite interesting after a few weeks.

On top of the chemise, working women might wear a sleeveless dress, with a gown on top. Her chemise and dress or skirt would generally be worn above the ankle, to avoid trailing in the mud or in the dust indoors. Rich women normally wore boned corsets to give a flat-chested, long-waisted look, with bodices sewn to very full skirts, and several petticoats underneath. In order to make their skirts balloon outwards they often wore elaborate farthingales – frameworks of whalebone hoops worked into cloth, usually of the petticoat. In winter they might wear knee-length drawers for warmth. On their legs they usually wore knee-high stockings, held up with garters. When Mary Queen of Scots was led out to her execution on 8 February 1587, she was wearing a pair of sea-blue socks on top of white Jersey hose. (Jersey was originally wool or worsted, but later any knitted fabric. Making this material was already an important industry on the island of Jersey.)

*A smock dating from 1575 which Tudor women wore as an undergarment beneath their dress or skirt.*

A man's underwear consisted of a long shirt, rather like the chemise, and perhaps underpants. On top of this he wore doublet and jerkin. The jerkin was a short sleeveless jacket worn over the doublet, which was a waisted tunic that came down to mid-thigh, with sleeves that were sometimes puffed. On his legs he wore hose – short or long stockings, or sometimes tights – and knee-breeches. The hose were always tight, but the breeches might be either tight or immensely baggy and highly elaborate. On top of all this he generally wore a cloak and a hat.

One male accessory that really came into prominence in Tudor times was the codpiece, named after the old English word 'cod' meaning scrotum ('not in polite use' – OED). Breeches and hose were made with slits so that the wearers could urinate without taking all their clothes off, but because buttons were unreliable, and zips had not been invented, they needed something to preserve their modesty. At first the codpiece was just a simple triangular flap of cloth covering the slit, but as doublets became fashionably shorter the codpiece came into view, and men realized they had a new way of showing off their virility. They began to wear immense padded codpieces, sometimes in strident colours, and often decorated with ribbons and jewelled pins.

Various reasons have been suggested for why Henry VIII displayed impressive codpieces in middle age. One was that he suffered from syphilis – which may have explained some of his symptoms – and that the codpiece was necessary to hide bandages and ointment. Another was that he wanted to show the world that even though he had no male heir there was nothing wrong with his sexuality. At any rate, the codpiece saw its finest hour in the 1540s, and gradually disappeared from view during Elizabeth's reign.

However, the shortness of the doublet did mean that warm stockings were popular in the winter, for men as well as for women. According to legend, Queen Elizabeth was presented in the second year of her reign with a pair of black silk stockings by her silk woman Miss Montague, and was so delighted she said she would never wear woollen ones again. Certainly she was given various pairs of silk stockings, and no doubt giving such a present was good for your court rating.

# Mechanization

Elizabeth certainly wouldn't have worn the stockings made by an early piece of labour-saving machinery. It was invented in 1589 by the vicar of Calverton, a village just north of Nottingham; his name was William Lee. According to (Victorian) legend he was madly in love with a young woman,

*A Tudor nobleman
displays his
sumptuous costume.*

*A model of the stocking frame invented by Reverend William Lee in 1589.*

but she continually put him off because she was always knitting; so he designed and built a machine to do the knitting so that he could have her to himself. Another story is that he loved his wife dearly, and wanted to release her from the drudgery of knitting his socks.

Whatever prompted him, Lee took a break from his religious duties and invented the stocking frame, which was intended to speed up the knitting process by using many needles instead of just two. The yarn was pulled into loops by 'sinkers' that moved up and down, and the loops were caught by ingenious 'bearded' needles to create the coarse knitted material. This was ideal for the stockings that men wore with their short breeches.

Lee's machine had eight needles to the inch, and produced 600 loops a minute. Today's computer-assisted machines have many more needles, and produce five million loops a minute – and yet the principle remains broadly the same as Lee's original idea.

Lee took his stocking frame to London – which must have been quite a journey – and won an audience with Queen Elizabeth, to demonstrate it. She was underwhelmed to see the coarse wool stockings it produced; by this time she was wearing those fine silk stockings. So he went back home, built an improved machine with twenty needles to the inch, and in 1598 began to knit silk stockings fit for a queen. By this time, however, his sponsor Lord Hunsdon had died. Hunsdon was also his access to the queen, and so he never did get a chance to show Elizabeth what his improved machine could do. Furthermore the queen was worried about putting hand knitters out of work, and therefore not keen to endorse Lee's machine; as a result she would not grant him a patent.

In due course he was persuaded to go to France by the promises of rich rewards from King Henry IV, and he set up a factory in Rouen with his brother and nine workmen; his apprentices showed their pride in the craft by wearing their silver working needles on chains around their necks. But yet again Lee's luck ran out. In 1610 Henry IV was murdered, and without his royal support Lee could not carry on: he died in Paris.

His brother James Lee brought seven of the apprentices back to England and carried on the business, improving, enlarging and promoting the machinery. Slowly, knitting frames began to appear in English homes, and by the eighteenth century there were thousands of stocking frames in use; they had become the basis of an important new industry, especially in the east midlands, where Lee had invented the idea. This was one of the first stirrings of the Industrial Revolution – the tide of machines that would come to dominate life for the Victorians.

# Shop till you drop

One of the places where Lee's stockings would have been sold was the Royal Exchange, built by wheeler-dealer Sir Thomas Gresham, who had acquired a fortune by sometimes dubious deals in Antwerp, where he was Royal Agent for both Edward VI and Mary, and returned to England in 1567. All the dealing in what amounted to the stock exchange was then done in Lombard Street in the city of London, and the traffic had to be excluded because of all the men haggling in the mud. Gresham wanted to create a building like the Bourse in Antwerp, and built what he thought was a suitable structure on Threadneedle Street.

The ground floor was designed as the trading floor, but he realized this would not make him much money; so he built two more floors above, with about a hundred retail outlets, or kiosks. Each one was to pay him an annual rent, which eventually rose to seven, eight or ten pounds. To begin with, he had difficulty filling the kiosks, but when news came that in January 1570 the Queen was going to visit he persuaded the existing tenants to fill all the other shops with goods, and said they could keep them rent-free for the rest of the year. He then gave Her Majesty a fine dinner before showing her round. She was much impressed, and said it should be called not the Bourse but the Royal Exchange.

After the queen's visit the Royal Exchange suddenly became fashionable, and a wonderful range of merchants congregated in the country's first shopping mall, from haberdashers, mercers (silk dealers), and milliners to armourers, apothecaries and barber–surgeons; and you could buy anything from a gold chain to a mousetrap.

The queen was concerned that trading should be fair, not only in the new Royal Exchange but throughout the country. One vital component of fair trading was standardized weights and measures. Her grandfather Henry VII had issued a few standard weights, but she decided to do the job thoroughly. Accordingly in 1574 she set up a committee to examine the problem and advise her, and after much argument – because there were two sets of weights in general use – they settled on new standards. In 1588 fifty-seven sets of standard weights were made and sent to the mayors of every town in the country, so that disputes between traders could be settled. These included flat-round weights, bell-shaped weights, and 'cup' weights that nested together. One master set remained in the Tower of London. These weights remained as standards until the 1820s. Elizabeth later issued standards for the pint, quart and gallon.

*The Royal Exchange, so named by Queen Elizabeth in 1571.*

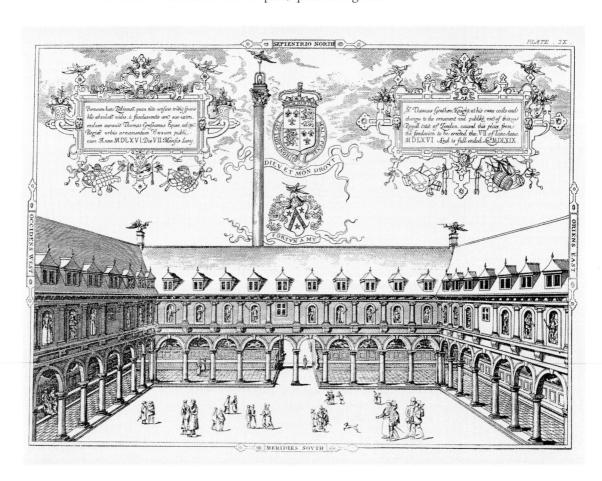

The Royal Exchange burned down in the great fire of 1666, was rebuilt but burned down again in 1838. The current majestic building is Victorian, and higher than the original, although it still boasts on a tower a fine golden grasshopper, emblem of the Gresham family. Curiously, the Royal Exchange has rarely fulfilled its promise. The dealers never thronged to its dealing floor, and the merchants stayed away in droves from the retail outlets.

*These Tudor weights remained as standards until the 1820s.*

So what did the Tudors do for us in the way of pleasure and the good life? Better food – including potatoes, tomatoes and the sweet course – better beer, better music, more games, more gambling, the tennis racket, the shopping mall, machine-made stockings, standard weights and measures, wallpaper, the theatre, the incomparable Shakespeare and his illustrious contemporaries, a load of proverbs, and some other memorable rhymes – like this one from Richard Grafton's *Abridgement of the Chronicles of England* (1568), which counted off the months of the year whatever your station in life:

*Thirty days hath September*
*April, June, and November;*
*All the rest have thirty-one,*
*Excepting February alone,*
*And that has twenty-eight days clear*
*And twenty-nine in each leap year.*

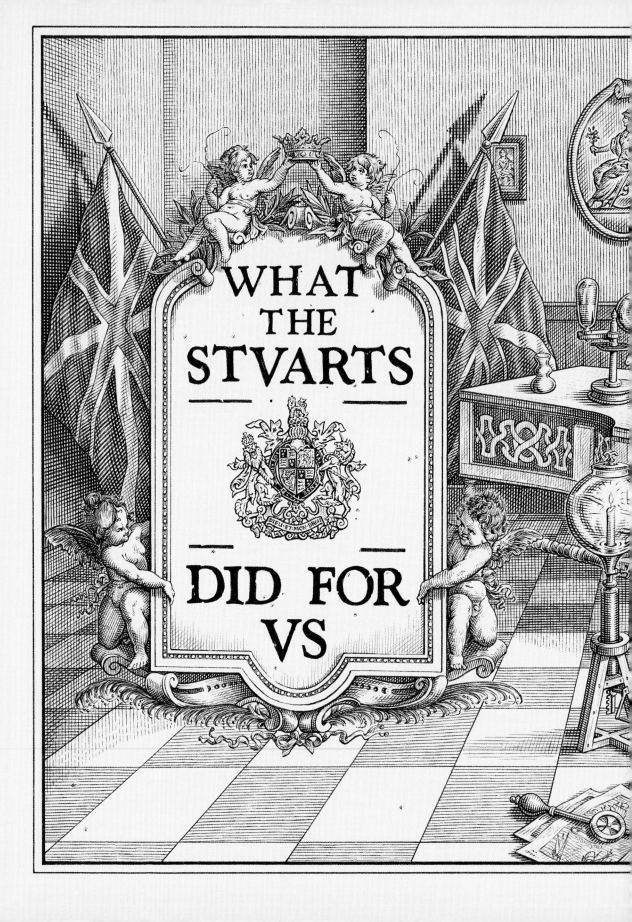

# The Stuart dynasty

James I became King of England in 1603. He was already King of Scotland, through his ill-fated mother Mary Queen of Scots. While by no means unintelligent, he alienated his English subjects by his extravagance and his favourites (a succession of handsome young men). Crucially, he believed in the concept of divine right: he was king by God's decree – not a stance that would be popular with Parliament.

James had married Anne of Denmark and they had three children, the eldest of whom was Henry, Prince of Wales. Henry was bright, handsome and popular; but he died young. His brother Charles had been a sickly child, but he survived and became king in 1625; he married Henrietta Maria of France. Charles's reign was ended by the Civil War, the most momentous event since Henry VIII's split with Rome. Charles was even more intransigent than his father on the question of divine right; essentially he wanted to be a dictator in a country where the idea of democracy was growing, and he repeatedly dissolved

The Royall Oak.

Parliament. Actual hostilities began in 1642, and civil war raged for seven years until Charles was tried in Westminster Hall, found guilty of high treason and executed in Whitehall on 30 January 1649.

After his death, Oliver Cromwell, the Parliamentary leader, declared Britain a republic: the Commonwealth. He ruled as Lord Protector until 1658; when he died, the fledgling republic died with him. Not only was there no fundamental unity in Parliament; traditional royalist feelings were growing stronger too. By 1660 the dead king's son Charles was able to return to his country in triumph.

Charles II was a shrewd, cynical politician, a man of great energy and a patron of art and science. He was renowned for his many mistresses, and fathered at least fourteen bastards. But his wife, Catherine of Braganza, was barren; when Charles died he was succeeded by his brother James II.

As a Catholic, James was unpopular in the country. He had two daughters by his first wife, Anne Hyde, after whose death he married the Catholic Princess Mary of Modena. He was tolerated while his elder daughter Mary, a Protestant, was the heir; but then in 1688 his son was born, to great controversy – for years the rumour was that a baby had been smuggled into the room in a warming pan. The country dreaded a Catholic succession, and Mary's Protestant Dutch husband, William of Orange, was invited by Parliament to take over. In the Glorious Revolution of 1688, he duly invaded, and he and Mary became joint rulers. James fled to France; attempts to reinstate his son (the Old Pretender) or his grandson (the Young Pretender – Bonnie Prince Charlie) were fruitless, and eventually his line died out.

Mary died in 1694, William in 1702; they had no children, so they were succeeded by Mary's sister Anne, popularly known as 'Brandy Nan'. She married Prince George of Denmark, and had no fewer than eighteen pregnancies. Tragically, only one child, William, survived – and even he died at the age of eleven. Anne's final years were melancholy; she died in 1714 (by which time she was so fat that her coffin was almost square).

So once again there were no direct heirs; the throne passed to the great-grandson of James I: the Protestant Elector of Hanover, George I. The Stuarts had replaced the Tudors, and now the Hanoverians replaced the Stuarts.

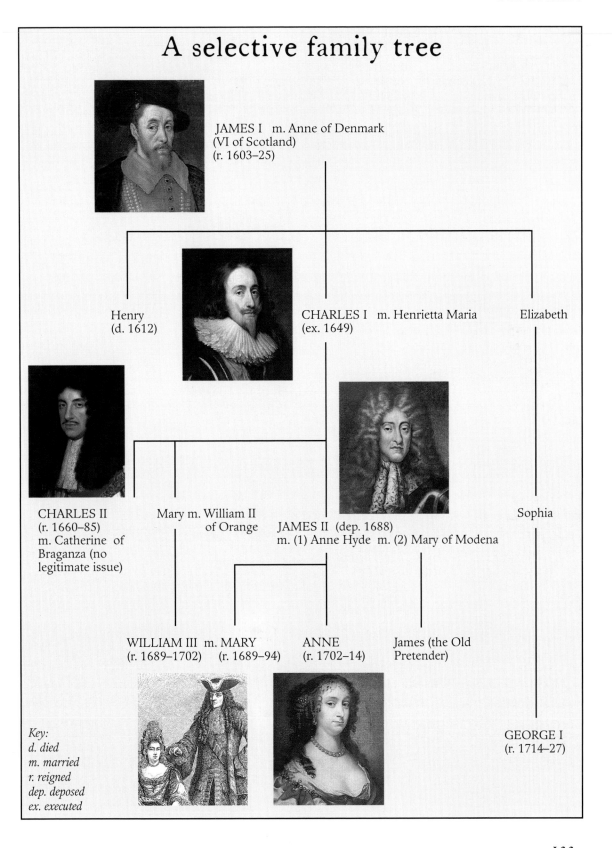

# A selective family tree

JAMES I   m. Anne of Denmark
(VI of Scotland)
(r. 1603–25)

Henry
(d. 1612)

CHARLES I   m. Henrietta Maria
(ex. 1649)

Elizabeth

CHARLES II
(r. 1660–85)
m. Catherine of
Braganza (no
legitimate issue)

Mary m. William II
of Orange

JAMES II  (dep. 1688)
m. (1) Anne Hyde  m. (2) Mary of Modena

Sophia

WILLIAM III  m. MARY
(r. 1689–1702)   (r. 1689–94)

ANNE
(r. 1702–14)

James (the Old
Pretender)

GEORGE I
(r. 1714–27)

Key:
d. died
m. married
r. reigned
dep. deposed
ex. executed

## Being the Fifth Chapter

# THE ORGANIZED ISLE

The Restauration of Monarchy
KING CHARLES II.

'Infinite the crowd of people and the
horsemen, citizens, and noblemen of all
sorts... The shouting and the joy expressed
by all is past imagination.'

SAMUEL PEPYS DESCRIBES THE ARRIVAL OF CHARLES II

*Previous spread: Charles II, restored to the throne in 1660, was an enthusiastic lover of fun, entertainment and women.*

he seventeenth century was one of the most tumultuous in England's history. Long-running religious tensions boiled over, and the split between king and Parliament resulted in not only the cataclysmic Civil War, but – unimaginably – the execution of an English monarch. When James I came to the throne in 1603, he had promised to be friendlier to Catholics after Elizabeth's reign, but in fact their persecution continued. There was huge resentment at this, a resentment that found its ultimate expression in the Gunpowder Plot – an event that has printed itself on the national psyche. As the well-known story goes, a group of conspirators led by Robert Catesby planned to blow up Parliament and the king; the plot was discovered, along with Guy Fawkes and the barrels of gunpowder. The plotters paid the ultimate penalty, and their failure is commemorated even now, nearly 400 years later, with the annual Bonfire Night when Guy Fawkes is burnt in effigy.

*Guido Fawkes' signatures before and after torture.*

*The signature of Guido (Guy) Fawkes before and after he was tortured on the rack.*

The passing centuries have crystallized the Gunpowder Plot in those rather simplistic terms; there is reason to believe, however, that Stuart spin doctors – primarily Robert Cecil, Earl of Salisbury and the king's Chief Minister – discovered the plot early on (or perhaps even fabricated it in the first place) and manipulated events to their advantage. We will probably never know the full truth.

The legacy of that aborted plot may be burning a guy once a year, but other developments in Stuart times have had a rather more substantial effect on the country. This was a time when scientific genius flowered, and technology took off – later chapters will deal with many of these aspects. The Stuarts also began to make improvements to the quality of life, tackling the problems of water supply, public transport and street lighting, introducing a public postal service and publishing the first proper newspapers; commercial life benefited from the organization of banking, and agriculture from a crucial invention. But first came a symbolic change on the flagpole.

# The organized flag

When the Scottish King James came to the throne of England, he realized that he needed a symbol of unity between the two countries. Now that they shared a king, what flag should they fly to celebrate the union? The English flew the flag of St George, a simple red cross on a white background. The Scots flew the flag of St Andrew, a diagonal white cross on a blue background. The solution was to combine the two, but that led to some

*The first union flag representing England and Scotland.*

argument. The Scottish favoured slapping their flag on top of the English one, but the English protested vigorously, and in the end the compromise was as shown here. The Scots complained that their flag was underneath; the English complained that their white background was almost flattened by the Scottish blue, and by the time the first Union Flag was agreed on 12 April 1606, it had already caused international disagreement.

The flag was popularly called the Union Jack (and still is); there are a number of theories why. The word 'jack' was used as a diminutive, and a small flag was often flown from the small mast on the stern of a ship; or the flag may have been named after the small mast – the jackstaff. When the red cross of St Patrick was added for Ireland in 1801, the Union Jack became even more complicated, assuming the design familiar today.

# The Authorized Bible

One of the first things James did when he came to the English throne was to support a call for the Bible to be revised – indeed, 'authorized'. William Tyndale's translation (see page 23) had been followed by further revisions, notably the Geneva Bible of 1560 which proved popular. A number of churchmen had long believed that existing translations were corrupt; James himself had a low opinion of the Geneva Bible and expressed a wish for a 'uniform translation'.

Accordingly, the king appointed a number of learned men – around fifty of them – to translate the original Greek and Hebrew from scratch, but also with reference to existing translations. These men,

deemed the best biblical scholars and linguists of the time, laboured for years; their work was finished by 1611, and the title page of this new Holy Bible set out the basis of their endeavour:

Conteyning the Old Testament
AND THE NEW
Newly Translated out of the originall
tongues & with the former Translations
diligently compared and revised by his
Majesties Speciall Commandment.

Appointed to be read in Churches.

The Authorized Version was to eclipse all previous versions and become the 'standard' both at home and abroad. It remains the bestselling book in the world, acclaimed for the beauty and clarity of its language, which has suffused our culture with quotations and proverbs – many of them due to William Tyndale.

# Organized water

Under the Tudors, the city of London had grown considerably – and grown in a haphazard way. Houses were built chaotically and, with no sewage system, the city was filthy and disease-ridden. Crucially, there was no effective water supply, without which improvements in hygiene would be impossible. This was recognized, and in 1605 and 1606 Acts of Parliament were passed to try and improve the supply of water. In 1607 King James himself issued a challenge to any man who could devise a workable system, but nothing practical was done. Most people still depended on water either from wells or from the river; since 1582 a great waterwheel at London Bridge had pumped water out of the Thames. This supplied thousands of customers, but the water was not exactly pure, even a hundred years later: after a shower in the city in 1710, Jonathan Swift wrote of the torrent running down the streets:

*Sweepings from Butchers' Stalls, Dung, Guts, and Blood,*
*Drown'd Puppies, stinking Sprats, all drench'd in Mud,*
*Dead Cats and Turnip-Tops come tumbling down the Flood.*

The man who rose to James's challenge was Hugh Myddelton (or Middleton), born in Wales, the sixth son of the governor of Denbigh Castle. Myddelton went off to London and became a goldsmith and banker. He also started trading by sea, encouraged by his friendship with Sir Walter Ralegh; they used to sit together in the doorway of Myddelton's shop and smoke the new-fangled tobacco.

In 1609 Myddelton offered to construct a London water supply himself, and on 21 April began work on what came to be called the 'New River'. In four and a half years, despite strenuous objections from the owners of the land he had to cross, he dug 40 miles of canal about 10 feet wide and 3 feet deep, from springs near Ware in Hertfordshire to a reservoir called the New River Head in Islington, in north London. Paying off the landowners was so expensive that he ran out of capital, but the king was impressed by the work in progress, and agreed to pay half the total cost, in exchange for half the profits.

The opening of the New River was celebrated with a great public ceremony, hosted by the Lord Mayor of London, who was Myddelton's elder brother Sir Thomas Myddelton. It must have been a fine family occasion, and in 1622 Hugh Myddelton was made a baronet by the king.

From the reservoir, water was taken into the city in pipes made from hollowed logs of elm, jammed together end to end. Water was then carried into individual houses by lead pipes in the basement, and by 1670 many areas of London were well served, as many as two houses in three having their own water supply. Even today some parts of Stoke Newington in north London are still supplied by the New River.

*Old London Bridge as it appeared in 1600 from the south-east. From 1582 a great waterwheel supplied thousands of Londoners with water.*

# Organized transport

Another area crying out for attention was public transport, which at the beginning of the seventeenth century was almost non-existent. If you wanted to risk your life, you could take a ferry across or along the Thames in London, but the watermen were notoriously rude, the river was effectively an open sewer, and the boats occasionally sank. Rich people had horses to ride and carriages to be carried in, but everyone else had to walk. All that was to change in 1643, when a Captain Baily, who had once sailed in the fleet of Sir Walter Ralegh, launched a fleet of his own.

## Taxi!

Captain Baily bought four horse-drawn carriages, set up a table of standard fares, and employed uniformed drivers. Their instructions were to wait for customers in a rank in The Strand, next to a big maypole: the London taxi had arrived.

The carriages were pulled by huge ambling nags, known in Flemish as *hacquenées*, and so became known as hackney carriages. The fares were standardized at 6d (2½p) per mile for journeys up to 6 miles, and for longer journeys £1 a day for a four-horse carriage or 10 shillings (50p) a day for a two-horse carriage. Captain Baily's idea was such a success that it was soon taken up by dozens of other coach owners, and at any time there could be twenty cabs vying for trade. By the end of the century licences had been issued for 700 hackney carriages in London. Today there are some 20,000.

However, there were some unseemly incidents. In 1694 a group of women, masked and in party mood (clearly forerunners of ladettes), hired a coach and charged around Hyde Park. They behaved so disgracefully that cabs were completely banned from the park – and they weren't allowed back in until 1924.

## Lighting the way

It would have been difficult enough for hackney carriages to negotiate London's crowded, narrow streets during the day, but even worse by night. There was no lighting, so the streets would be pitch dark, also making it hazardous for any pedestrian venturing out. Then in the mid 1680s Edmund Heming was granted a patent giving him the exclusive right to light the streets. His plan was to light an oil lamp by every tenth house on every moonless night from Michaelmas (29 September) until Lady Day (25 March). He started with Kensington Road. He also offered to provide lights for stables, mines, yards, and coaches that travelled late at night.

Other people soon realized that this was a great idea, and started muscling in. Worried that his workmen would be bribed by his enemies, he tended all his lights himself, both at midnight and in the early hours of the morning – which came close to killing him. He never became rich, since rivals stole his idea without paying him, but he was the first to provide street lighting. Not that his lights were universally popular. They didn't throw out much light, since he had to encase them with sheets of horn rather than glass, and they burned smelly fish-oil, which tended to drip on to the heads of passers-by.

## Long-distance travel

So while you could at least hail a cab in London to get you to your destination, outside the city (unless you rode a horse) the only form of transport was the coach. There weren't many of them in the early 1600s, and most of those on the roads were private. However, in the summer of 1657

*Travelling by stage coach was exceedingly slow and uncomfortable in the 1600s.*

'My journey was noe way pleasant. This travell hath soe indisposed mee, that I am resolved never to ride up againe in ye coatche. I am extremely hot and feverish.'
EDWARD PARKER

*One problem with regular coach services was that the highwaymen knew where and when to lie in wait.*

three enterprising men organized a regular coach service from London to Chester. According to their advertisements, you could get on the coach at Aldersgate on Mondays, Wednesdays and Fridays, and travel to Coventry in two days for 25 shillings (£1.25), to Stone in three days for 30 shillings (£1.50), and to Chester in four days for 35 shillings (£1.75).

Today the journey from London (Aldgate or Victoria Station) to Chester by coach (with a lot of horsepower but no horses) takes a little over seven hours and costs £18.50, which is cheap compared with the fare in 1675.

The journey in those days was not for the faint-hearted. The coach was uncomfortable and smelly; you were jammed in and hardly able to move. The wooden benches were hard, your feet were in muddy straw. The roads were terrible (see below), and so the ride was extremely rough; you might easily get sick. There was no heating or cooling – apart from the open windows – so that you would freeze in winter and cook in summer. A vicar called Edward Parker complained bitterly after a journey from Preston to London in 1663: 'My journey was noe way pleasant. This travell hath soe indisposed mee, that I am resolved never to ride up againe in ye coatche. I am extremely hot and feverish.'

As if the fever-inducing ride was not enough, the traveller also had to worry about highwaymen. Once the stage-coaches had adopted regular timetables the robbers knew where and when to lie in wait, and therefore became a worse menace than before. One of the most notorious was a highwaywoman, Moll 'Cutpurse' Firth, who was said to be foul-mouthed, incredibly ugly, and to dress like a man.

She became infamous in her own lifetime, and had a play written about her, called *The Roaring Girl*. The stories told are probably exaggerated, and possibly quite untrue, but it was said that she smoked a pipe and kept weird pets, including a psychotic dog called 'Wildbrat' that went everywhere with her. She started life as a pickpocket, and became highly skilled at it before turning to highway robbery.

She even managed to hold up and rob General 'Black Tom' Fairfax, the ruthless commander of the Roundhead Army, and one of the most feared men in England. The hold-up took place on Hounslow Heath; he had a guard of armed men, but she shot two of them dead. She was caught at Turnham Green, but bought her freedom with £2,000 of the money she had stolen. Later she became a fence, dealing in stolen goods from a pub of ill-repute two doors from the Globe in Fleet Street. She died at seventy-five, a rich woman, from natural causes, and asked to be buried face down in order to continue her rebellion against the Establishment.

*Moll 'Cutpurse' Firth, notorious and successful highwaywoman.*

Even today transport presents a nest of difficulties, and no doubt carjacking and motorways are debated frequently in Whitehall. In the seventeenth century it occupied some of the great scientific minds; Christopher Wren and Robert Hooke spent at least one whole afternoon discussing transport at the Royal Society (of which more in later chapters). The worst problem then was the state of the roads. They had hardly been touched since the Romans had left more than a thousand years before; so they were rutted and bumpy, pot-holed dirt tracks, quagmires in winter and rocklike in summer. In 1663 the Turnpike Trusts began, with the aim of collecting tolls from road users and spending the money on repairs. However, they had little impact for a hundred years.

While the roads remained terrible, the best hope seemed to be to improve the technology of the vehicles so that they could cope with the dreadful surface. On 1 May 1665 the diarist Samuel Pepys witnessed a trial of experimental coaches, all designed to make travel more comfortable. He was most impressed with one of them: 'Several we tryd, but one dyd prove mighty easy… Ye whole body of that coach lyes upon one long springe. It is very fine, and likely to take.'

He must have been describing suspension, which surely made life a whole lot better for the passengers.

# Maps and inns

One advantage of a less bumpy ride was that you could read your new map. A street map of London had appeared in 1559, and then in 1617 Aron Rathburne and Roger Burges took advantage of a new law, passed by James I. Before then, patents had been handed out rather indiscriminately by kings and queens, giving all sorts of monopolies away without any proper records being kept. James decided that the system should be subject to simple rules, and that all patents should be written down and recorded. The first patent under the new system, Patent No 1 of 1617, was granted to Rathburne and Burges for the production of street maps. The patent is in the form of a letter from the king saying that his beloved subjects Rathburne and Burges can make city maps – or, more precisely, that 'they onelie, and none others, may... make, describe, carve, and grave... all such maps... not only of London and Westminster, but also of York, Bristole, Norwich, Canterburie, Bath, Oxford, Cambridge, and Windsor'. And King James himself signed it, on 11 March 1617.

The sad thing is that we don't know whether Rathburne and Burges ever did produce their equivalent of A–Zs, since the possession of a patent does not force you to make anything. However, we do know that the country's first proper road atlas, *Britannia*, was published in 1675. It had 100 separate engravings of roads running in strips up the page, with notes about the surrounding countryside, and it covered 7,500 miles of accurately measured roads; indeed, it set the standard for road maps to this day. There had been a good deal of argument about how long a mile should be, but *Britannia* simply declared that the correct distance was London's statute mile of 1,760 yards – and that was that; the mile has been 1,760 yards ever since.

The man responsible for *Britannia* was John Ogilby, an extraordinary man who was born in Edinburgh in 1600. He became a dancing teacher (until he broke his leg), a calligrapher, a soldier, and was then nearly blown up and completely shipwrecked during the Civil War. He walked to Cambridge, learned Latin, translated Virgil, learned Greek, translated Homer, built a theatre in Dublin, finally settled in London to print books, and after being appointed a 'sworn viewer' or surveyor of the disputed properties in the city after the Great Fire of 1666 (see page 187), he was given the wonderful title 'King's Cosmographer and Geographic Printer'. He produced beautiful books on China, Japan, Africa, America and Asia, and finally *Britannia*, at the express wish of the king.

He was also a dreadful poet, and among much other poor material he wrote three epic poems, of which, according to the *Dictionary of National Biography*, 'the first two were never published, and the third was fortunately burnt in the fire of London'. However, his road maps were excellent. Portable versions were printed – on the backs of playing cards.

'They onelie, and none others, may... make, describe, carve, and grave... all such maps...'

From Patent No. 1 of 1617

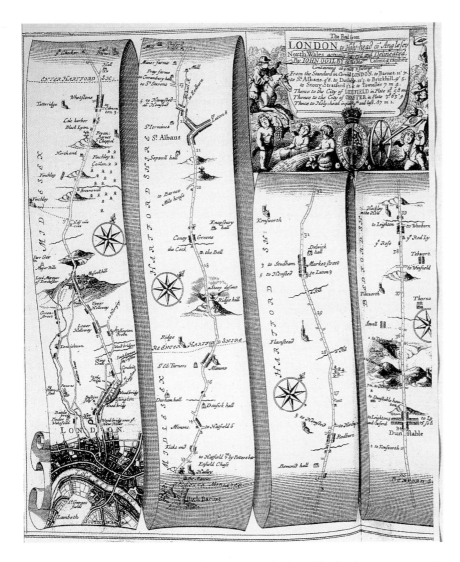

Card games were popular in the Stuart period, and in the late seventeenth century a fashion developed for playing cards that were 'educational' – that is, they were ordinary cards suitable for playing games, but they also carried educational information, about law, theology, heraldry, mathematics, spelling, Latin grammar, and especially geography. The theory was that the young people playing card games would at the same time acquire useful knowledge, and therefore not be completely wasting their time.

Around 1675 Henry Winstanley (see page 203) produced a set of playing cards with a geographical theme; each featured pictures of and comments about the countries and peoples of the world. Conventionally, the most important card was the king of hearts. Winstanley's king of hearts shows Charles II and Catherine of Braganza with, in the background, the Thames, the old London Bridge, and the monument designed by Christopher Wren and recently erected there to commemorate the Great Fire of London of 1666.

*Britannia, the country's first proper road atlas, was published in 1675. This page shows the first part of the road to Chester and Holyhead, the following page has notes on the landmarks.*

The potential market for such playing cards was extensive: at least sixty-five different educational sets of playing cards were produced between 1675 and 1720. Ogilby's maps were put on cards too; you could take them in your travelling case, amuse yourself and your companions on your journey, and at the same time make sure the coachman was going in the right direction.

Such informative cards would be on sale at the seventeenth-century equivalent of the motorway service station; that is, the coaching inn, which must have been a welcome sight for the tired and sore-bottomed travellers. However, the accommodation generally left quite a lot to be desired. No question of freshening up – no one bothered much with washing – and anyway a nip of brandy for fourpence was a more attractive choice. If you were early and lucky you might have been served a meal of, say, roast pigeon, and then if you tipped the landlord well enough you would be allowed to fight for a place in a large bed already full of strangers and fleas.

The important point about the inns was that they provided a change of horses; there were always horses resting there. Without this service the horses would have been permanently exhausted, and journeys would have been incredibly slow. Surprisingly, though, the coaching inns would never have existed at all if it had not been for another Stuart invention, the Post Office.

# Sorting the post

At the beginning of the seventeenth century the Royal Post was just one of the king's private privileges, set up by Henry VIII and enjoyed by subsequent monarchs. Teams of swift horses and messengers were available to carry letters to and from the kings and queens and their courts, wherever they were. It was essential to have staging posts, where horses could be exchanged and riders refreshed – hence the establishment of those coaching inns.

For ordinary people, however, the only way to send a letter was to take it yourself, try to bribe the King's Messenger, or find a tradesman who was travelling in the right direction and hope he would take it for you. Things changed in 1635, when Charles I decided to let his subjects use his Royal Post. Charles was not a popular monarch; despite seeing himself as God's Chosen One, he was corrupt and selfish and his obduracy precipitated the Civil War. Perhaps he was trying to curry favour with his restless subjects when he founded the 'Letter Office of England and Scotland' to provide a postal service – at a price. To make sure that no one encroached on his monopoly he issued a dire warning: 'His Majesty straightly charged and commanded all his loving Subjects whatsoever duly to observe his Royall pleasure therein declared, as they will answer the contrary at their perils.'

The Royal Mail was a great success, and twenty years later the General Post Office was born, and the first Postmaster General was appointed – just one more thing that the Stuarts did for us.

In 1680 William Dockwra established a penny postal system in London. He set up a receiving-house in each of the principal streets. Every hour the letters and parcels taken in at the receiving-houses were taken to sorting offices, sorted, registered and delivered all over London. There were four deliveries a day in the suburbs, and as many as eight in the City. All letters and small packages were delivered anywhere in the city for a penny, and to the suburbs for twopence. (Sadly, the post has gone into decline since then; now most of us are lucky to get two deliveries a day and, in spite of high-speed travel, the mail seems to move more and more slowly, bogged down by its own volume.)

Unfortunately there was resistance to Dockwra's scheme. City porters thought the post would take away their livelihood; and Titus Oates, a fanatical anti-Catholic conspirator and propagandist, denounced the postal service as a popish plot. The Duke of York (the later James II) was violently opposed to this enterprise, and as soon as Dockwra looked like making a profit he accused him of infringing the king's rights; Dockwra had to pay a hefty fine, and lost all his money. Nevertheless the scheme was the beginning of postage stamps as we know them today.

# Publishing and censorship

While people had long communicated by personal letter, the Stuart period saw a rise in publicly circulated information: in other words, news. In previous centuries, information was circulated among interested parties – merchants, say – by hand-written newsletters; the invention of printing with movable type in the fifteenth century (see page 18) brought the development of printed pamphlets, often for propaganda purposes. The first 'real' newspaper in English was *The Weekly Newes*, brought out in London in 1622. Even then it could report only foreign news: during the reigns of Elizabeth, James I and Charles I, all reporting of domestic politics had been forbidden. By order of Star Chamber (the court created by Henry VIII), 'all printing and publication of the same be accordingly supprest and inhibited'.

In fact royal censorship could be pretty touchy anyway, especially concerning anything seen as remotely criticizing the monarchy. One William Prynne fell foul of royal sensitivities on more than one occasion. He was a militant Puritan, crusading against anything he considered sinful in modern life. In 1632 he produced a thousand-page book attacking the theatre, claiming that plays were unlawful, condemned by the Bible, and incentives to immorality. In particular he condemned women who appeared in plays. Unfortunately in January 1633 Queen Henrietta Maria and her ladies took part in a play, and his comments were taken as a direct insult. What's more, Prynne's caustic comments about spectators and tyrants were taken as attacks

'His Majesty straightly charged and commanded all his loving Subjects whatsoever duly to observe his Royall pleasure therein declared, as they will answer the contrary at their perils.'
CHARLES I PROTECTS THE ROYAL MAIL

*Oliver Cromwell banned most things that were fun, even Christmas parties.*

on the king. He was tried in Star Chamber, imprisoned in the Tower for a year, then fined £5,000, had half of each ear cut off, and was sentenced to life imprisonment – and they took away his Oxford degree.

The first major step towards our current state of news saturation was in 1641, when Parliament abolished Star Chamber. The bill to free the press from royal censorship had been pushed through by the MP for Ely, the 41-year-old Oliver Cromwell. Now anyone could write anything. Immediately political news appeared, swiftly followed by propaganda on both sides. Newspapers began to appear two or three times a week, with a few real news stories. In what was to become an enduring aspect of the tabloid press, they carried some sensational reports, such as man-fish and mermaids being seen in the Thames (rather like Elvis being seen on the Moon); and to fill up the remaining space they also printed rumours, scandals and all sorts of abuse. Sir John Berkenhead, publisher of the pro-royalist *Mercurius Aulicus*, and one of the first writers to make a career out of journalism, was said by John Aubrey to 'lye damnably'; he referred to Cromwell as 'Carrot-nose' and as 'Beelzebub's Chief Ale Brewer'.

As well as newspapers, masses of pamphlets were being circulated, propagandizing whichever religious and/or political line the writers embraced. Parliament was getting scared of possible sedition, and in 1647 Cromwell, ironically, reintroduced censorship. He even paid spies to sniff out the sources of the underground press, but by then it was hard to stop. When Charles II came to the throne in 1660 he too tried to take control of the press by appointing a Surveyor of Imprimery in the Press Act of 1662, and in 1675 he became so worried about the coffee-houses (see page 194) – where the newspapers were not only read but also written – that he tried to close them down. Nevertheless, during his reign newspapers flourished. These included the first satirical paper, *Heraditus Ridens*, and the first anti-papal publications, such as the *Weekly Packet of Advice from Rome*, of which half was Christian news and the other half was salacious material. The *London Gazette* was first published in 1665, and is still with us today; it is probably the world's longest continuously published paper.

Censorship ended with the arrival of William and Mary, and the *London Spy* (1698–1700) was the first paper that covered only scandal. There were all sorts of gossip sheets, and in 1693 appeared the *Ladies' Mercury*, the first women's newspaper, which included the first agony column: 'All questions relating to love, etc are still desired to be sent in to the Latin Coffee House in Ave Mary Lane, to the Ladies Society there, and we promise they shall be weekly answered with all the Zeal and Softness becoming to the Sex. We likewise desire we may not be troubled with other Questions relating to Learning, Religion etc.'

On a rather more serious note, the *Tatler* (1709) embraced arts and politics, and claimed to tell its readers what to think, while the *Spectator* (1711) was full of essays and opinions. Daniel Defoe published a *Weekly Review* from 1704, and Jonathan Swift wrote for the *Examiner* from 1710 until 1714. Meanwhile the *Daily Courant* was opened by Elizabeth Mallett in March 1702, and was soon followed by other daily newspapers. A forum for informed public debate was emerging, with an eager audience; the daily – and the weekly – had arrived. So had journalists. Today many people have a low opinion of journalists, but this is nothing new; even in those inhibited Stuart days people distrusted them. The dramatist Ben Jonson wrote scathingly of 'Captain Hungry, who will write you a battle in any part of Europe in an hour's notice, and yet never set foot outside a tavern'. Similarly, the poet Richard Braithwaite wrote, 'thanks to his good invention he can collect much out of very little... here a city taken by force before it bee besieged; there a country laid waste before ever the enemie entered'.

'All questions relating to love, etc are still desired to be sent in to the Latin Coffee House in Ave Mary Lane, to the Ladies Society there, and we promise they shall be weekly answered with all the Zeal and Softness becoming to the Sex.'
THE LADIES' MERCURY

# The London Gazette.

## Published by Authority.

From Monday, February 5. to Thursday, February 8. 1665.

*Deal, Feb. 4.*

UPon the arrival here of two Ketches, within few houres one of another; the first reporting the Dutch to be before *Ostend* with 26, the other with 30 sail of Ships, Sir *Christopher Minnes* set sail this day about noon with the Squadron, under his Command, to the Eastward, of whom, we hope, in few days to hear more. A Ketch is now bringing in a light Prize, the next may give you a farther account of it.

*Yarmouth, Feb. 2.* This morning past by this Town, 20 sail of laden Colliers. We have advice of a Squadron of Dutch ships about the *Gun fleet*, of which, two are said to be Flag-ships.

*Harwich, Feb. 3.* This morning sailed by us Westward, several ships, great and small, supposed to be Colliers with a Convoy, who gave some salutes of Guns at *Oasley-Bay*.

*Plymouth, Feb. 2.* The *Antilope* is ready to set sail, and those of His Majesties Ships that suffered by the late storm, will be ready at farthest, in four or five days to go to Sea. We have advice, this Evening from *Dartmouth*, That the *Saphire*, and the *Forester*, have brought a Vessel into *Torbay*, of about 400 Tuns, 16 Guns, and 35 Men, whom we question not to be prize; the Pilot being Dutch, and the Master a Frenchman, laden from *Venice*, with Oyl, Soap, Rice, Sherry, &c.

*Lyn, Jan. 31.* A small Barke of this Town laden with Salt, arrived here this day, who came on Wednesday last out of *Tinmouth* Haven, without Company or Convoy, till off *Scarborough* he fell in with the *True-Love* Fregot, (suspected to be lost) and went under his Convoy as far as *Humber*.

*Amsterdam, Feb. 5.* The onely note here is Money. *Holland* hath called to the Provinces for the nine Millions towards the old Fleet of the 72 Ships, and for the two Millions for the building the new ones; and considering the great arrears *Zealand* is in, especially for the entertainment of the *Lunenburgh* forces, two Deputies are to be sent to quicken them in those payments.

The States of *East-Friesland* have again remonstrated the inconveniencies they suffer from the *Lunenburgh* forces, demanding to be freed from them, and offering to secure their Countrey themselves by other means. On the other side, the Princess Dowager presses as hard, that the Commander of *Lieroot* may keep certain Villages, she there mentions, in contribution; but yet she refuses to send any person to treat about the matter.

The Sieur *Colbert*, Envoy hither from the French King, is said to speak much for an accommodation with the Bishop, though the truth be, we begin to smel his design is not to see the Bishop ruined, but to keep up the Ball for his own ends; and that he intends to send the Count of *Furstenberg* under hand, to assure the Bishop, he means not wholly to have him undone. In the mean time, though we are glad to hear from *Paris*, that King is now for good and all declaring for us; yet we are a little startled at a hint we finde given to our Envoy there, that the King expects now we should agree about the supream command of the Fleet, when his is joyned to ours; a matter, we wonder any scruple should be made at, considering the rule that is given in the Treaty of 1635. by which, this is to be measured.

Five Companies are on the way from *France* hither, in exchange for the *Daulphins*, and d' *Estrade* hath been earnest for convenient Quarters to be forthwith assigned them, fearing otherwise how they will relissh the service, considering the disrepute it hath got in *France* from the hardships of the last Campagne.

Prince *Maurice* hath been desired forthwith to pass to *Wesel*, and that two or three Deputies of the States General, or of the Council of State, may accompany him with full power upon Communication with the Prince, to act as they shall see cause upon the place.

*Hague, Feb. 5.* The States of *Holland* are met, and among other points to be deliberated, that of the choice of a General, in all appearance will be the most considerable. *De Witte* at the sollicitation, as is supposed, of Monsieur *Colbert*, hath proposed Monsieur *de Turenne* to be the man, which his Collegues dislike; of whom, some would advance the Prince of *Tarante*, others, Monsieur *de Schomberg*, he that is in *Portugal*: But much the greater number of the States, and among them, *Zealand* more obstinately, stand firm for the Prince of *Orange* to be General, and Prince *Maurice* to be Mareshal of the Army. What the issue will be, time will shew.

The Bishop of *Munster*, we hear, hath distributed a considerable sum of money among his Army; in which it is certain, the common Troopers have received four Rixdollars a man: And it is said, he hath given out Commissions for 2x Regiments more, so as to have his Army of 30000 Men compleat next Spring.

*Munster, Jan. 22.* His Highness the Prince of *Munster* has been for some daies here very intent in perfecting his new Levies, and forming the Counsells of next Spring, having in the mean time placed his Troops in very good Garrisons; the Foot most in the Enemies Countries, and the Horse in his own Townes. We have reason to think here that the Dukes of *Lunenburgh* grow cold in their intentions to *Holland*, and are concerting how they may with best security to themselves, and decency to the World, break off their Contract of Succors with *Holland*. Towards which, it is said, they have resolved to send a fit person into *England*, to make their excuse to His Majesty of Great Brittaine, for what they have already done, justifying themselves upon their want of a more early knowledge, of the true grounds and design of this Princes Armes.

*Brussells, Feb. 5.* In consequence of the late Enterview held at *Tervers* between the Marquess *de Castel Rodrigo* our Governour, and the Count *de Furstenberg* Bishop of *Strasburg*, on the part of the Elector of *Colen*; We have advice here that the Duke of *Newburgh*, and the Elector of *Colen* are hastning their new Levies, in which the Duke of *Brunswick Hannover*, and several of their Neighbour Prince are said to joyn with them, resolving to preserve the Peace of the Empire according to the Treaty of *Munster*, and their late Alliance of the *Rhine*, against any Forraign Princes, that should upon any pretence whatever march into the Empire. Which, the French, it is said, are much surprised at.

*Genoua, Jan. 20. Centurioni's* Squadron of 3 Ships is arrived here from *Cadiz*, with a Million of peeces of Eight in money and barrs; the States Convoy which has been expected this moneth, it is said will bring a far greater Treasure. The Dutch here report 15 sayl of Ships to be coming out to reinforce them, and to unite with the French, who have called all their Merchants ships from these parts to *Toulon*, where they are to receive the Convoy of 16 sayle designed for the Westward.

*Seville, Jan. 27.* The Commissioners of *France* and *Spaine* are still upon the Frontier of *Handaye*, treating upon the Contentious River betwixt that and *Fontarabia*, which at last, it is feared will goe near to begin the quarrel betwixt those two Crownes.

Several Troops are marching toward the Neighborhood

B b

*The very first edition of* The London Gazette, *published in 1665.*

# Organized monarchy

Oliver Cromwell, who came to power almost by chance, ruled the country for eight years, but his puritanical regime gradually lost popularity. He banned almost everything that was fun – even Christmas parties – and, as we have seen, ended up censoring the press as much as the king had. When Cromwell died, the public celebrated, and demanded the return of the monarchy – but on different terms. Never again would Britain risk a monarch with absolute power. The new king would rule as a moral figurehead, while political power would stay in the hands of an elected government. And it was on these terms that Charles II was restored to the throne in the spring of 1660; this was the beginning of constitutional monarchy.

Charles II was then thirty years old, and had spent fourteen of those years in uncomfortable exile. He was received with enormous enthusiasm when he landed at Dover, as Samuel Pepys wrote in his diary: 'Infinite the crowd of people and the horsemen, citizens, and noblemen of all sorts... The shouting and the joy expressed by all is past imagination.' After a splendid procession via Canterbury to London – 'So glorious was the show with gold and silver, that we were not able to look at it, our eyes at last being so much overcome' – Charles found both houses of Parliament waiting to greet him.

While much given to the pursuit of pleasure (his many mistresses and bastards were no secret, and his popular nickname was 'Old Rowley' after a famous stud horse), he was also an enthusiastic patron of the arts and sciences. He was particularly interested in scientific developments, and it was during his reign that the Royal Society, the first scientific society, was chartered.

# Organized plants

One of the branches of scientific investigation, the study of plants, was formalized by the English naturalist John Ray. He was born on 29 November 1627, lived in Black Notley, near Braintree in Essex, and made it his life's work to collect and classify nearly 20,000 plants. His father was the village blacksmith and his mother the village wise-woman – a sort of Essex witchdoctor using herbal medicines; perhaps this was how he first became interested.

Ray became an academic, and taught Greek, humanities and mathematics at Cambridge, but in 1650 after a serious illness he began taking long walks as part of his convalescence. He became fascinated by the plants and flowers around Cambridge, and wrote a book describing more than 600 species: *Catalogus plantarum circa Cantabrigiam nascentium* – the first ever account of the flora of a county. But this merely whetted his appetite; with the help of

*John Ray, the incomparable botanist from Black Notley in Essex.*

one of his pupils, Francis Willughby, Ray set out to make a systematic description of the entire natural world.

Ray spent thirteen years on expeditions, collecting plant specimens and studying animals. Most plant books at the time (see Culpeper below) told readers whether they could eat a particular plant, or what illness it could cure. Ray took a totally different approach: he looked at plants as scientific specimens, and divided them into groups scientifically. He ignored superficial features like colour, or size, or where they grew. Instead he concentrated on their entire structure, and eventually decided he could divide plants into three basic types: plants with no flowers, plants with two seed-leaves, and plants with only one seed-leaf; botanists still use this classification today.

Ray produced book after book. His major work was *Historia Plantarum*, which was written in Latin and covers 18,600 species. He also wrote about fish, fossils, birds and mammals, and proverbs as well, and he even published a Greek, Latin and English dictionary. Fifty years later Carl Linnaeus, the great Swedish naturalist whose system of categorizing living things is still the world's standard, described him as 'The incomparable Botanist'.

# Organized health

Medicine in the Stuart period hardly qualified for the description 'scientific', still being a traditional mix of magic and superstition where, if a cure happened to work, it was more by luck than judgement. Professional practitioners – physicians, surgeons and apothecaries – guarded the secrets of their trade jealously; the College of Physicians was highly annoyed when in 1649 Nicholas Culpeper translated its *Pharmacopeia* (list of drugs) from Latin into English. The members felt their authority was undermined, once their dubious secrets were readily available; in fact Culpeper's 'crime' was reminiscent of the 'heresy' of William Tyndale in publishing the Bible in English, a hundred years earlier (see page 23).

Born in London in 1616, Culpeper set himself up as a physician and astrologer in Spitalfields. He was well respected as a doctor, and was spurred into translating the *Pharmacopeia* by those of his rivals who not only charged their patients enormous fees for doing very little, but did not even understand what they were talking about.

Fortunately for Culpeper, the College of Physicians did not have legal authority, and had to content themselves with mudslinging; they described his book as:

> done (very filthily) into English by one Nicholas Culpeper… an absolute
> Atheist, and by two yeeres drunken labour hath Gallimawfred the
> apothecaries book into nonsense, mixing every receipt [i.e. recipe] therein
> with some scruples, at least, of rebellion or atheisme, besides the danger
> of poysoning men's bodies. And (to supply his drunkenness and leachery
> with a thirty shilling reward) endeavoured to bring into obloquy the
> famous societies of apothecaries and chyrurgeons.

Undeterred, Culpeper went on to write a string of medical books, of which the best known was *The English Physician Enlarged*, with 369 medicines made of

*Nicholas Culpeper's* Herbal *was an enormously popular self-help medical book.*

*English herbs that were not in any impression until this*. These books were immensely popular, and new editions appeared in succession; *The English Physician Enlarged*, also known as *The Herbal*, was still being reissued in the early 1800s. Perhaps people were impressed by the length of the full title, which was *The English Physician or an Astro-physical Discourse on the Vulgar herbs of this Nation, Being a Compleat Method of Physick Whereby a Man may preserve his body in Health, or Cure himself being Sick, for threepence charge, with such Things as onlie Grow in England, they being Most Fit for English Bodies*. So what Culpeper did for us was to provide a manual that gave ordinary people more control of their health, and diminished the power of the pseudo-medical fraternity.

## Pollution

We can only guess how effective Culpeper's nostrums really were; certainly in Stuart times the population was assailed by all manner of infections and pollutants. A particularly noxious irritant in the city was the smoke produced by thousands of fires, domestic and industrial. The diarist John Evelyn wrote of 'the horrid Smoake which obscures our Churches and makes our Palaces look old, which fouls our Clothes and corrupts our Waters'. Evelyn was commissioned by Charles II to write a pamphlet about air pollution, which he entitled *Fumifungium: or the Inconvenience of the Aer and Smoake of London Dissipated*. Evelyn pleaded with the king and Parliament to cut down the burning of coal in London: 'and what is all this, but that Hellish and dismall Cloud of SEACOALE... so universally mixed with the otherwise wholesome and excellent Aer, that her Inhabitants breathe nothing but an impure and thick Mist accompanied with a fuliginous and filthy vapour...'

Coal was often known as sea-coal because it was collected from beaches, where seams of coal had been exposed by erosion (it was also often transported by sea). The wonderful word 'fuliginous' means sooty; Evelyn wrote in his diary on 24 January 1684: 'London... was so filled with the fuliginous steam of the sea-coal, that hardly could one see across the streets.' Sadly coal-burning went on being a problem in London until 1952, when the terrible smog of early December killed 2,000 people, and forced the government to act. After the Clean Air Act of 1956 the 'pea-souper' or 'London Particular' gradually ceased to be a problem. So the Stuarts at least started the ball rolling in the long fight against pollution of the air we breathe.

# Organized finance

As trade increased during the Stuart period so also did the need for organized finance. People could not reasonably carry with them enough gold to do big deals in the city, and some form of recognized credit had to be set up. To begin with, this was done by goldsmiths, who in effect doubled as bankers. Then in 1691 a rich, influential merchant called William Paterson proposed the founding of a Bank of England. Paterson managed to persuade William III that this was a good idea, because William's home country, Holland was at war with France, putting a massive strain on his finances, and the king needed money urgently. Paterson approached 1500 rich people with the scheme, promising to make them more money. He raised £1.2 million, set up the Bank, and subsequently lent the money to the government at eight per cent interest; so the shareholders got a good return for their investment. The loan was only supposed to last for twelve years, but somehow the government never got around to paying it back. Our national economy today rests on that hastily-created loan made three hundred years ago; it's called the national debt. With the Bank of England came the first banknotes, which were originally receipts for each of those first deposits. However, the critical thing about the new banknotes was that they said 'I promise to pay the bearer on demand the sum of ten pounds [or whatever]' – which they still say today. This made them a reliable form of exchange (although today you would attract some funny looks if you were to walk into any bank and demand your money in gold…). In effect these notes were imaginary money, and doubled the amount in circulation.

The Bank of England has functioned ever since, but Paterson himself did not always pick such reliable ventures. In 1695 he formed the Scottish Africa and India Company, which was basically an attempt to set up a colony in

*The first banknotes were essentially just receipts for money deposited.*

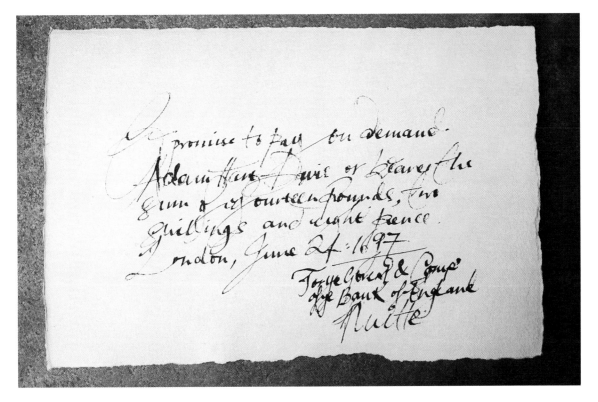

Darien, in eastern Panama, in order to control trade between the eastern and western worlds. This aroused much excitement, because of Paterson's standing and persuasive powers, and he managed to raise almost a million pounds, much of it from Scots. However, the entire scheme collapsed, and everyone lost their money.

# Organized agriculture

Trade may have been opening up, and banking undergoing a revolution, but one type of activity hardly seemed ripe for change: farming. People have been farming for millennia – probably over 9,000 years. However, even in agriculture there is evolution, and the first proper agricultural machine was invented by a Stuart, Jethro Tull, who was born in 1674. At the time, seeds were planted by hand (or 'drilled') into ploughed furrows. Farmers sowed as many seeds as possible in their land, in rows close together. Tull believed he could produce more crops by sowing fewer seeds; he insisted that his workers follow strict guidelines, planting precisely spaced seeds in rows well apart. His workers were furious – how dare this pen-pusher tell them how to do their job! They refused to co-operate, and went on strike.

*Our reconstruction of Jethro Tull's seed drill.*

Tull decided that if he couldn't make his men do the work, he would have to invent a machine to do the job instead. While he was a student, he had been fascinated with musical instruments, and especially with the organ. He remembered how an organ lets air into a pipe with a sprung flap, and he designed a horse-drawn machine that would drop seeds into a pipe in the same way. He called it a seed drill.

Although at the time most farmers laughed at his ideas, by the end of the Stuart period it was impossible to ignore his success; he managed to grow wheat on the same patch of land for thirteen years, without ever using manure. There was such demand for his methods that eventually he published a book, *The New Horse Hoeing Husbandry*. This was the start of the agricultural revolution that would gather pace in the middle of the eighteenth century. And a modern seed drill does essentially the same job as Tull's original.

But the revolution nearly didn't happen. During one nasty scene with his indignant workers, Tull had to call in the army, who shot several of his men. In his frustrated rage he smashed up his machine, and threw the pieces down a well – and as far as anyone knows, they are still down there today.

So what the Stuarts organized for us was a Union Flag and a definitive Bible, a water supply for London, the first public transport both in London and across the country, the turnpike laws and street lighting, the first public postal system and the first real road maps, the first newspapers, the first herbal remedies printed in English for anyone to read, the first botanical classification of plants, and the first agricultural machinery.

## Being the Sixth Chapter

# NEW WORLDS

'Many shall sail through, and
knowledge shall be increased.'

FROM FRANCIS BACON'S 'NOVUM ORGANUM'

he intellectual revolution that swept through Europe in the sixteenth century began the process of challenging long-established orthodoxies; for the first time, as we saw in Chapter 1, authorities revered for centuries were being questioned in many fields of human belief. This iconoclasm gathered pace in the seventeenth century, particularly in the field of scientific investigation; here, it was aided by unprecedented advances in optical instruments – especially microscopes and telescopes – which opened the eyes of the Stuarts to whole new worlds: the tiny and the cosmically huge.

Scientific developments were underpinned by a philosophy that rejected the 'wisdom of the ancients' as a given – these new worlds were seen with new eyes. One man embodied this modern approach: Francis Bacon.

# A new natural philosophy

Born in 1561, Bacon went to Trinity College, Cambridge, and was immensely successful. He became a lawyer and a Member of Parliament, held a number of important public posts including that of Lord Chancellor, and was created Viscount St Albans and first Baron Verulam (the Roman name for St Albans was Verulamium). Then it all went horribly wrong; he was arrested in 1621 on a technical charge of bribery, and briefly imprisoned in the Tower of London, after which he effectively retired from public life and spent his time in philosophy and writing.

His important philosophical books were *The Advancement of Learning* (1605), *Novum Organum* (1620) and *The New Atlantis* (published in 1627, the year after his death). He developed the notion that scientific knowledge and understanding could be gained by experiments, and that improved understanding would lead directly to benefits for the human race. *Novum Organum*, written in Latin, was so called because it was a new (*novum*) approach to the acquisition of knowledge, which in Greek Aristotle had called *Organon*. The frontispiece of *Novum Organum* showed the Pillars of Hercules, symbolizing the limits of ancient human knowledge, and a ship sailing through into the modern world that Bacon dreamed about, with a Latin inscription that meant 'Many shall sail through, and knowledge shall be increased.'

Aristotle had said that the way to reach the scientific truth about anything was by observation, authority and argument. If you had clever enough men, and if they argued for long enough, they were bound to reach the truth. Bacon disagreed sharply, and said that the Aristotelians were like spiders, spinning webs out of their own substance. The way to find the truth, he said, was to interact with the real world, to investigate, and to do experiments. He may have been influenced by Galileo, who had shown

> **'Whether or no anything can be known, can be settled not by arguing, but by trying.'**
>
> FRANCIS BACON

dramatically what observation and experiment could do (see page 44), but in any case Bacon's philosophy was clear: 'Whether or no anything can be known, can be settled not by arguing, but by trying.' In other words, scientific truth should be based on evidence rather than imagination.

*Novum Organum* has a collection of about 140 aphorisms, which are long-winded but penetrating statements about the nature of objective knowledge. Aphorism XLV from Book 1 begins: 'The human understanding, from its peculiar nature, easily supposes a greater degree of order and equality in things than it really finds; and although many things in nature be [unique] and most irregular, yet will invent parallels and relatives and conjugates, where no such thing is. Hence the fiction, that all celestial bodies move in perfect circles...'

*Francis Bacon, Viscount St Albans and first Baron Verulam: the philosopher who advocated experimental science.*

> 'There is much ground for hoping that there are still laid up in the womb of Nature many secrets of excellent use.'
>
> FRANCIS BACON

Aphorism XLVI begins: 'The human understanding, when any proposition has been once laid down... forces everything else to add fresh support and confirmation; and although most cogent and abundant instances may exist to the contrary, yet either does not observe or despises them, or gets rid of and rejects them by some distinction, with violent and injurious prejudice, rather than sacrifice the authority of its first conclusions.' He backs up this idea of cognitive dissonance with a vision of a man being shown in a temple a picture of people saved from shipwreck after they had prayed for deliverance, and asked whether that did not prove the power of God. 'Ah,' says the man, 'and where are the pictures of those who were drowned?'

All this is essentially theoretical, but in *The New Atlantis* he conjures up a vision of a hands-on science centre, the House of Salomon, which would contain 'Perspective-houses, where we make demonstrations of all lights and radiations... We procure means of seeing objects afar off, as in the heaven and remote places... We have also engine-houses... Also fire works for pleasure and use... we have ships and boats for going under water... We imitate also motions of living creatures, by images of men, beasts, birds, fishes, and serpents.'

He also imagines a college that will initiate scientific exploration; thirty-six fellows will be divided into groups, each with a specific area of research, to 'establish the causes of things, and amass such a collection of facts as would lead to new discoveries and inventions'. And he spells out what is in effect the driving force for most scientists: 'There is much ground for hoping that there are still laid up in the womb of Nature many secrets of excellent use. They too, no doubt, will some time or other come to light of themselves, just as the others did; only by the method of which we are now treating [i.e. the experimental method] they can be speedily and suddenly and simultaneously presented and anticipated.'

Bacon inspired a real-life college of like-minded searchers after truth – the Royal Society, which was to number among its members such giants as Robert Boyle, Robert Hooke, Christopher Wren and Isaac Newton, of whom more later. The Society began as a series of ad hoc discussions during the 1640s (despite the distractions of the Civil War), held sometimes in Oxford and sometimes in London; without a fixed base, the London group called themselves the 'Invisible College'. King Charles had a keen interest in scientific matters and took a great interest in their work; soon after his restoration the Invisible College took on more corporeal substance when in 1662 it was chartered as the Royal Society of London for Improving Natural Knowledge.

Sadly, Bacon's enthusiasm for hands-on science proved fatal. He wondered whether meat might be better preserved if it were kept cold, and in April 1626 he bought a chicken on Highgate Hill and stuffed it with snow that was lying on the ground. We do not know how long the bird stayed fresh, but Bacon caught a chill and died, in the very process of inventing the frozen chicken.

One of the earliest scientists to work along the lines advocated by Bacon was William Harvey, the man who discovered the circulation of the blood.

# The world of the blood

Probably born in Folkestone in 1578, Harvey was a most distinguished doctor. He became chief physician at St Bartholomew's Hospital, physician extraordinary to James I, and physician ordinary to Charles I, in which capacity he looked after the royal princes at the Battle of Edgehill, allegedly sitting under a hedge reading a book. He may have been the first person in England to become addicted to coffee, but his main claim to fame is what he worked out about the circulation of the blood.

He studied at Cambridge, where he caught malaria in 1599, but the following year he was well enough to go off to Padua University, the best medical school in the world at the time. Harvey was a student there while one of the professors was the great Galileo, and the place seems to have been rather a nest of radicals. Even though the superb anatomy work of Vesalius had been published there more than fifty years earlier (see page 36), and Padua boasted the finest anatomy theatre in the world, the medical teaching still came from the works of Galen, published 1,400 years before.

*William Harvey depicted here in a life-size portrait on a wall panel at the University of Padua.*

One of Harvey's professors was the famous Fabricius de Aquapendente, who was interested in the valves in human veins – clearly visible as lumps on the back of many people's hands and wrists. In fact, he often gets the credit for discovering them, although in fact they had been known for some time. He thought they were there to slow down the flow of the blood.

The current theory (that is, Galen's) was that there were two sorts of circulation – venous in the veins and arterial in the arteries. Venous blood was made in the liver in order to nourish the body. It constantly ebbed and flowed, getting replenished by seeping through the pores in the 'septum' wall from the right of the heart to the left, but in general it flowed out into the body during the day, and back to the heart at night. The arterial circulation carried air from the lungs, which carried away the heat from the

heart – although Galen must have seen blood spurting from the arteries of wounded gladiators.

Harvey studied the human body, and investigated living animals, and came to the conclusion that Galen's theory was rubbish. The arteries contained blood, not air, and the valves in the veins completely stopped the flow downwards. The only function of the veins must be to carry blood up to the heart. Having dissected many living creatures, he was sure the heart was a pump, and reckoned that in humans it pumps about 60 ml (say, a quarter of a cupful) every beat, which is about once a second. It follows that in half an hour the heart must pump about 100 litres of blood – enough to fill a bath. Clearly the liver cannot possibly make all this blood, and in any case there is not enough space for it in the body. Instead, the same relatively small amount of blood must be going round and round the body.

He reasoned that the heart must be a pump with two chambers. The left side sends blood around the body through the arteries. The blood comes back through the veins to the right side of the heart, from where it is sent off to the lungs to be refreshed (in modern terms to dump carbon dioxide and pick up oxygen), ready for the next cycle.

Harvey first explored these ideas in lectures from about 1615, and in 1628 he finally published his great book, *Exercitatio Anatomica de Motu Cordis et Sanguinis in Animalibus:* or *Anatomical Essay on the Motion of the Heart and Blood in Animals.* This was a vast step forward for physiology, and in particular one of the first applications of the modern scientific method. Instead of simply believing what the ancients had said, Harvey applied first observation and then experiment to try to come up with a more convincing explanation of what was going on: he was one of the pioneers of modern science.

The only hole in his theory – or rather the only absence of holes – was at the extremities. How did the blood in the fingertips, say, get from the arteries into the veins? The question was finally answered by Marcello Malpighi in Bologna in 1661; he examined the lungs of a frog, and clearly saw with his microscope the tiny tubes (the capillaries) that carried the blood from the arteries through the muscle.

So Harvey's ideas, controversial at the time, were eventually supported by microscopical evidence – which was underpinning all kinds of exciting discoveries.

# Microscopic worlds

One of the men involved in the nascent Royal Society, Robert Hooke, published the first high-quality pictures of microscopic objects. Born on the Isle of Wight in 1635, he was a fascinating and brilliant, but irritating, man. He went to Oxford, where he was assistant to Robert Boyle, for whom he constructed an air pump (see page 223), and possibly helped with the mathematics for Boyle's Law. Hooke became Curator of Experiments and then Secretary at the Royal Society, and remained there for fifty years.

Like many of his illustrious contemporaries, his genius was many-sided – not for these pioneers the one-track path of specialization. He was enormously ingenious; he claimed to have invented the anchor escapement mechanism for a clock, soon after Huygens made a pendulum clock (see page 211), and also in 1658 he invented the watch spring. In 1675 Hooke had a watch made for Charles II inscribed *Robert Hook inven. 1658. T. Tompion fecit 1675*. He is remembered in Hooke's Law, which explains the extension of elastic springs. He was the first to describe cells, the building blocks of all living things. He may have invented the universal joint. However, he created problems for himself; whenever anyone else came up with an idea or an invention, Hooke had the maddening habit of saying, 'Oh yes, I thought of that five years ago,' or, 'I made one of those, but it worked better...' This did not endear him to people. With his rivals he could be peevish and cantankerous.

In 1665 Hooke published an extraordinary and ground-breaking book, *Micrographia*. This contained some telescopic observations of the Moon and

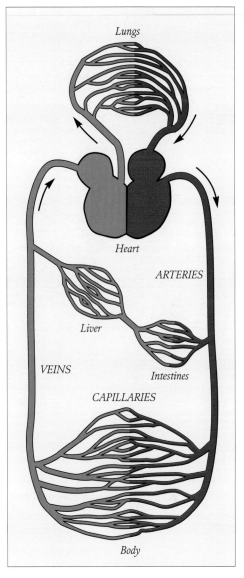

*William Harvey was the first to describe the circulation of the blood.*

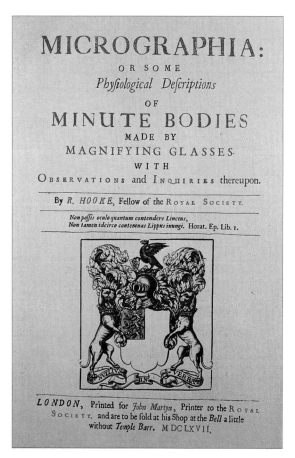

MICROGRAPHIA:

OR SOME

*Physiological Descriptions*

OF

MINUTE BODIES

MADE BY

MAGNIFYING GLASSES.

WITH

OBSERVATIONS and INQUIRIES thereupon.

By R. HOOKE, Fellow of the ROYAL SOCIETY.

*Non possis oculo quantum contendere Lincens,*
*Non tamen idcirco contemnas Lippus inungi.* Horat. Ep. Lib. 1.

LONDON, Printed for *John Martyn*, Printer to the ROYAL SOCIETY, and are to be sold at his Shop at the *Bell* a little without *Temple Barr.* M DC LXVII.

*Title page of Robert Hooke's ground-breaking book* Micrographia, *published in 1665.*

stars, descriptions of a thermometer, in which Hooke defines zero as the temperature of freezing water, and discussion of the difference between heat and burning. He suggests that fossils have an organic origin. Above all, though, he describes his microscope and the things he has seen through it. His drawing of a flea is dramatic even today, and in 1665 it must have been sensational. He also drew or described lice, weevils and other little beasts, and the structure of cork, which he said seemed to consist of many small cavities – he called them 'cells' – all joined together. We now know that all living things are made of such cells, but Hooke was the first to see them and name them.

Hooke did not invent the microscope; the first seems to have been made by the Dutchmen Hans and Zacharias Janssen at Middelberg in 1590, and brought to England in 1617 by Cornelius Drebbel (whose own claim to fame was to construct the 'boats for going under water' imagined by Bacon – see page 162). However, Hooke's meticulous observations and published drawings were a revelation.

One man who was directly inspired by Hooke's book, and became as a result a great pioneer of renaissance science – in fact he could be called the father of microbiology – was Antoni van Leeuwenhoek. His father had been a basket-maker, and he set up as a linen draper, in his home town of Delft, where one of his friends was the artist Jan Vermeer. He was accustomed to use a magnifying glass to examine the quality of his cloth, and when on a trip to London he saw a copy of *Micrographia* – and in particular Hooke's pictures of taffeta and linen – he was inspired to try making and using his own microscope.

He began in 1671, and ground his own lenses from tiny pieces of glass. Imagine a glass ball – or a round glass vase full of water. Hold an object close behind the glass ball and you can see a magnified image through the ball. Van Leeuwenhoek realized that an infinitely large glass ball would give no magnification, that a ball the size of his fist might give a magnification of say two times, and therefore that a very small glass ball should give much higher magnification.

He was secretive about his methods, but somehow he managed to make lenses that gave him a magnification of about 300 times. He took a sheet of brass around the size of a playing card and drilled a small hole in the centre.

In this hole he mounted his tiny glass lens, and this was his microscope. He mounted his subject near to the lens and moved his eye close up to the other side of the lens. While the brass sheet cut out all the stray light, he moved his eye to and fro until he could see a greatly magnified image of his subject – and he entered a whole new world.

Van Leeuwenhoek found he could see things that no one had ever seen before. He looked at bee stings, lice, fleas and fungi and, excited by what he saw, he wrote to the Royal Society in London, describing his discoveries. The members were so incredulous – even Hooke must have been doubtful – they wrote back and asked for proof; they suspected he might be making it all up. He repeated some of his observations in front of a priest and got a signed witness statement supporting his claims.

In 1674 he looked through his microscope at a drop of the slimy green water from a nearby lake, and was amazed to see that it was full of tiny plants and animals – the first micro-organisms that anyone had ever seen. He wrote:

*Above: A replica of van Leeuwenhoek's microscope. Below: Robert Hooke's drawing of a flea, from Micrographia, contemporary viewers would never have seen objects magnified in this way before.*

> *On examining this water… I found floating therein divers earthy particles, and some green streaks, spirally wound serpent-wise… about the thickness of a hair of one's head… there were, besides, very many*

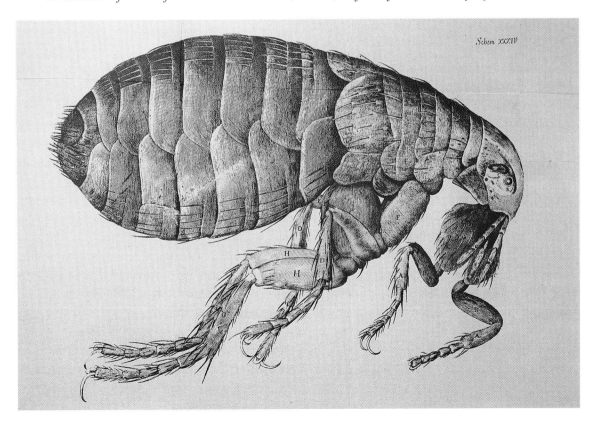

Schem XXXIV

*little animalcules, whereof some were roundish, while others, a bit bigger, consisted of an oval. On these last I saw two little legs near the head, and two little fins at the hindmost end of the body… And the motion of most of these animalcules in the water was so swift, and so various upwards, downwards, and round about, that 'twas wonderful to see: and I judge that some of these little creatures were above a thousand times smaller than the smallest ones I have ever yet seen, upon the rind of cheese, in wheaten flour, mould, and the like.*

IV. *An Extract of a Letter from Mr.* Anth. Van Leuwenhoek, *Concerning Animalcules found on the Teeth ; of the Scaleyneß of the Skin,* &c.

I Have often endeavoured to discover Animalcules in Spittle, but in vain: But examining a kind of gritty Matter from between my Teeth, and mixing it sometimes with Rain-water, and sometimes with Spittle, both which before had no Animalcules, I discovered therein with admiration a great number of very small ones moving ; the greatest whereof are represented Fig. 2. *A.* they had a very strong and swift Motion in the Water like Eeles, of these larger there were not many : A second sort is represented by *B.* these oft turned themselves round like a Top, and moved sometimes as is shewn by the Line *C. B.* these were much more in number. The Figure of the third sort I could not well discover, sometimes they appeared oval, and at other times round, they were so small that I could not discern them greater than as at *E.* they moved swiftly by each other like Gnats playing in the Air, and of these I discerned Thousands in a drop of Water they shewed no bigger than a Sand, and in the Drop, the Water was to the Animalcules as 9 to 1. But most of the Matter I examined, consisted of long slender parts all of a thickness, but differing in length as at *F.* and one crooked one amongst the rest ; and because I have formerly observed Animalcules of this shape in Water, I endeavoured to discover if these lived but could not. I have found the same in the Matter taken from between the Teeth of other

*An extract from one of Antoni van Leeuwenhoek's numerous letters to the Royal Society, describing his observations of the plaque from his teeth.*

His reports of this new world of nature beyond the limits of unaided human vision were exciting, but also terrifying to the philosophers of the day. They had enough trouble describing the normal world, and to be confronted with untold numbers of new kinds of living things was an unpleasant shock.

In the end, during more than fifty years, van Leeuwenhoek wrote 190 long letters to the Royal Society. His favourite observation was of the blood flow in eels; he built a special eel-holding microscope. He was the first person in the world to see spermatozoa and, most amazingly, bacteria. Sadly, while he was studying a disease of sheep, he fell ill himself, and died of what is now called van Leeuwenhoek's disease.

# Worlds beyond Earth

In the 1540s, as we have seen in Chapter 1, Nicolaus Copernicus began to challenge the old cosmology, which placed the Earth in the centre of the universe. The early years of the seventeenth century saw the telescope – aided by the powerful tool of mathematics – turned on the heavens, and astronomy would never be the same again.

Like the microscope, the telescope was a Dutch invention – or at least that is the recorded history, although there are hints that Leonard Digges (page 87) and Thomas Harriot (page 102) may have made telescopes earlier. In any case, in 1609 Hans Lippershey had the idea of putting two lenses together to make a telescope. Lippershey was a maker of spectacles, and the story goes that children were playing with lenses in his shop, and found they could make the church spire look bigger by putting two of them together. Lippershey constructed a simple telescope; the news of his invention travelled rapidly, soon reaching the radical Galileo in Padua.

Galileo made his own telescope, and immediately worked out how to improve it.

Lippershey's instrument magnified about three times. By experimenting with the geometry of the instrument and the curvature of the lenses, Galileo eventually achieved a magnification of about thirty times. He sold the idea to the army, and made himself some money, but then he turned his instrument towards the heavens, and the world was changed for ever.

The first thing he looked at was the Moon, in the autumn of 1609, and he was astonished. Aristotle had said that all heavenly objects were perfect spheres. Furthermore, while everything on Earth was made of the four elements – earth, air, fire and water – and was therefore changeable, all in the heavens was made of a fifth, immutable, element, and could never change. When Galileo turned his telescope on the Moon he saw at once that it was not a perfect sphere. The inner edge of the crescent moon was ragged and irregular, and within the dark area were many small points of light.

The only explanation he could find was that the surface of the Moon must be covered with mountains; the bright pinpoints were the mountain tops just lit by the rising or setting sun. Whatever the explanation, however, Aristotle had clearly been wrong.

Galileo was amazed again when he looked at a patch of fairly clear sky. Where with the naked eye he had seen just a few stars he could now see

*Galileo observed that the moon was not perfectly smooth or spherical.*

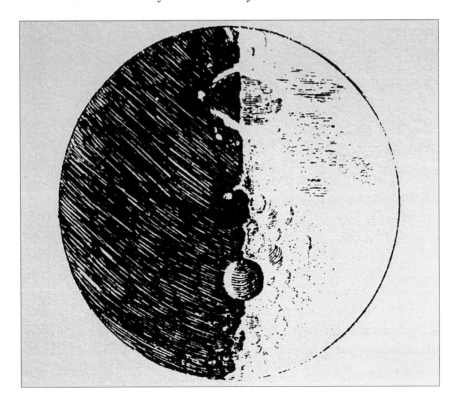

dozens: his telescope 'set distinctly before the eyes other stars in myriads which have never been seen before, and which surpass the old, previously known, stars in number more than ten times'.

*Jupiter and its biggest moons as observed by Galileo on 7 January 1610 (top) and the following night (above).*

But the biggest shock came when he looked for the second time at the planet Jupiter, one of the brightest objects in the night sky. When he first looked, on 7 January 1610, he noticed three bright points of light in line with the big disc of the planet – one to the left and two to the right – and assumed they were distant stars. However, the following night he looked again, and Jupiter seemed to have moved against these 'stars', for all three were to the left of the planet! Thinking he must have made a mistake, he waited with bated breath to look again on the 9th – but it was cloudy and he couldn't see a thing.

Eventually he realized that the bright 'stars' were not stars at all, but little satellites – moons revolving around the giant planet. He was able to see four of them, although often one or two were hidden by being in front of the planet or behind it. We now know that Jupiter has at least sixteen moons, but the ones Galileo saw are the four biggest: now called Ganymede, Io, Europa and Callisto. You can see them yourself with a small telescope or good binoculars, as long as you can hold them steady.

The theories of Copernicus had not cut much ice with most people, who continued to believe that the Earth was the centre of the universe, and that everything revolved around us. The moons revolving around Jupiter proved at a stroke that this was wrong. The Earth could not be the centre of the universe, and surely Copernicus must be right. The Professor of Philosophy at Padua refused even to look through Galileo's telescope, afraid of what he might see. This ocean of new stars was highly disturbing; it seemed not only to diminish the importance of our world, but to challenge Aristotle and therefore to upset the whole structure of existence.

Galileo took his telescope to Rome and showed his instrument and Jupiter's moons to the grandees in the Vatican. He was received in triumph, given an audience with the Pope and a ceremonial banquet, and his instrument was christened with the Greek word 'telescope'. They congratulated him then, but later became most upset when he openly supported the Copernican theory, and said that the Earth moved around the Sun. He ridiculed the idea that the Earth was at the centre of the universe, but he could not prove that the Earth moves. He was forced to recant – to declare publicly that he did not believe that the Earth was not the centre of the universe. Legend has it that under his breath he said, 'But it still moves!' In any case, poor Galileo was put under house arrest, and spent the last fifteen years of his life in his villa at Arcetri, just outside Florence.

IOANNIS KEPPLERI
Mathematici Cæsarei
hanc Imaginem

ARGENTORATENSI BIBLIOTHECÆ
Confecr.

MATTHIAS BERNEGGERVS
Kal. Ianuar. Anno Chr.
M DC. XXVII

# The worlds become clearer

When Johannes Kepler, a German professor of mathematics, heard of Galileo's observations of the moons of Jupiter, he immediately wrote him a long letter of support, and later, when he had got hold of a suitable telescope, he made his own observations and published them.

Kepler had been invited to Prague by Tycho Brahe (see page 32), and became his assistant there until Brahe died in 1601. Most of Kepler's life was spent in fruitless astrology, and in trying to construct a mathematical model of the universe that was in tune with the geometry of the regular platonic solids, or the harmony of the spheres. However, in between all this he did manage to apply intelligent mathematics to Brahe's superb

*The dreamy and mystical mathematician, Johannes Kepler, who worked out three laws of planetary motion.*

observations, and as a result was able to make accurate analysis of the movements of the planets. He eventually worked out three 'laws' that described planetary motion:

*The planets move round the Sun not in circles but in ellipses. The Sun is at one focus of the ellipse.*

*Planets move more quickly when they are closer to the Sun, in such a way that an imaginary line drawn from the planet to the Sun sweeps out equal areas in equal times.*

*The outer planets move much more slowly than the inner ones. For each planet, the time for one revolution is proportional to the cube of its distance from the Sun.*

Now that telescopes were available, anyone could look at the stars, see new things, and try to explain what they saw. Furthermore, because of the mathematical advances, knowledge of the stars began to have more practical uses. In England, two of the first people to explore this new ocean of information were Jeremiah Horrocks and William Gascoigne.

Horrocks was born near Liverpool around 1619, and managed to get to Cambridge, but found no one there to teach him about his passions, mathematics and astronomy; so he had to learn on his own. He bought a telescope, and also constructed some simple but elegant instruments.

He wanted to find out whether Kepler was right in saying that the Moon went round the Earth in an ellipse rather than a circle. If so, he reasoned, then the Moon takes a month to go round the Earth, and it must be closer to the Earth at some time during the month and further away at another. Therefore it should look larger when it is closer to us. He made a small hole in a card and held it on a stick at arm's length from his face. He looked at the Moon through the hole, and moved the card towards himself and away again until the Moon just fitted into the hole in the card. Then he noted how far the card was from his eye. By repeating this process night after night and comparing the results, he showed that the Moon does indeed appear to get larger and smaller during its cycle, and thus added support to Kepler's theory.

In 1639 he went to work as a bible clerk at St Michael's church at Much Hoole, between Preston and Southport, and it was there that he made his great discovery. Poring over some astronomical tables, he spotted what looked like a mistake.

About every 130 years, the planet Venus passes between the Sun and the Earth; in fact it does so twice about eight years apart. As Venus goes in front of the Sun, it can be seen as a little black spot, moving over the surface. This

'We may by their means make navigations as well into ye Heavens and discover new Countries there, as Columbus did by ships in America.'

HENRY OLDENBURG ON TELESCOPES

Erce gratissimum spectaculum et tot votorum materien

*A memorial window in St Michael's church at Much Hoole, showing Jeremiah Horrocks observing the transit of Venus.*

'Transit of Venus' was of great interest, because in principle, if it can be observed with precision from several places on Earth, then the observations can be used to measure the distance between the Earth and the Sun, which no one had been able to do.

Kepler had predicted that after 1631 the next transit of Venus would be in 1761, but Horrocks reckoned there should also be one in 1639, and when he refined his calculations he predicted it would be on 24 November, at about 3 pm. Luckily he discovered this in early October – six weeks ahead – and wrote to his brother Jonas, and to a fellow astronomer, William Crabtree, and asked them to watch out for the event. The day before he set up his telescope ready, just in case he had the date wrong.

Unfortunately 24 November was a Sunday; and Horrocks had to attend to his duties through several long services in the church. Each time he rushed back to his room to look for the transit, and at 3.15 pm he was rewarded: the clouds cleared, and he 'beheld a most agreeable spectacle' –

a little black disc moving slowly across the circle of the glowing Sun. He reckoned the diameter of the disc was about thirty times less than that of the Sun – much smaller than Kepler had suggested. At 3.50 pm the Sun set, which meant that Horrocks was unable to watch the entire transit, but he made as many observations as he could, and these were immensely useful for those who observed the next transits, in 1761 and 1769. The clouds covered the Sun for his brother Jonas, but Crabtree also saw the transit; he and Horrocks were the only people to do so.

As a result of this brilliant piece of prediction and observation, Horrocks has been called the 'father of English astronomy'; the later English astronomer Sir John Herschel called him 'the pride and boast of British astronomy'. Sadly, he died young, in early 1641.

Over in Leeds, his friend William Gascoigne had been as intrigued as Horrocks by Kepler's observations, and thought of a cunning way to test his ellipse theory mathematically. Just like Horrocks, he realized that if the Moon moved in an ellipse, then it must be closer to the Earth at some times during each month than at others, and it should look bigger when it was closer. He wanted to measure exactly how much its apparent size varied, and see whether it fitted his mathematical predictions.

He invented a micrometer screw – essentially a pair of callipers that he could adjust with great precision – and mounted it in his telescope so that the jaws were in the focal plane. Then he looked at the Moon, and adjusted his micrometer until the jaws just appeared to touch each side of the silver disc. This was like Horrocks's method, but much more precise, because he was able to make accurate measurements of the apparent size of the Moon. He repeated his observations night after night and found, as he expected, that the Moon did indeed seem to get bigger and smaller during the month, and by amounts that closely matched the theory.

His work was acclaimed: John Flamsteed (see page 177) was immensely impressed, and called Gascoigne 'the most ingenious man who ever lived', while the Scottish mathematician David Gregory said, ''Twas a pity he died so young [in his early thirties, fighting in the Civil War], when many naughty fellows live till eighty.' Christopher Wren was also impressed by Gascoigne's micrometer, and developed it further when he wanted to make an accurate globe of the Moon.

## The redoubtable Wren

Christopher Wren is renowned today mainly for his rebuilding work in London after the Great Fire in 1666 (see page 191). In addition to his masterpiece, St Paul's Cathedral, he built fifty-one churches in London and a great range of other buildings; he was the defining architect at the end of the seventeenth century. This reputation has eclipsed his endeavours

> '...the divine felicity of his genius combined with the sweet humanity of his disposition – formerly, as a boy, a prodigy; now, as a man, a miracle, nay, even something superhuman.'
>
> A DESCRIPTION OF CHRISTOPHER WREN

in many other spheres – for Wren too was a remarkable polymath, and made his own contribution to the growing science of astronomy.

Born in East Knoyle in Wiltshire in 1632, Wren went to Wadham College, Oxford, when he was seventeen, and stayed on at the university, becoming a Fellow of All Souls College in 1653. He seems to have been extraordinarily creative during the 1650s: as well as astronomy, he was interested in anatomy, lens-grinding, ciphers, fortifications, military engines, a double-writing instrument, a 'Strainer of the Breath' (to remove the carbon dioxide and make the same air breathable again), a machine to 'weave many Ribbons at once with only turning a Wheel', water-pumps, whale-fishing, a weather-clock, new musical instruments, a speaking machine and new surveying techniques.

He also carried out a number of grisly experiments removing the spleens from dogs, to see whether they survived (some of them did). The idea was to disprove Galen's theory of the four humours (see page 37), for without a spleen the dog could have no black bile, and the humours must be seriously if not fatally unbalanced. Certainly a demonstration of Bacon's advocacy of hand-on research…

In 1657 Wren was appointed Professor of Astronomy at Gresham College, founded in London in the late 1500s by the Tudor entrepreneur Sir Thomas Gresham. This meant that the new professor had to spend a good deal of time in London; as a member of the 'Invisible College' (see page 162), he could provide a base for meetings in his rooms there. The Oxford group were keen to understand the universe, and one of them, Henry Oldenburg, wrote that by improving telescopes, 'We may by their means make navigations as well into ye Heavens and discover new Countries there, as Columbus did by ships in America.'

King Charles was interested in science, and in May 1661 was invited to Gresham College to look through a telescope at Jupiter's moons and the rings of Saturn. The king was well aware of the fascination of astronomy, and of its fundamental importance, both in unravelling the nature of the universe and, more practically, in navigation. Any competent sailor could measure latitude – how far north or south he was – by measuring the elevation of the sun at noon, as long as he knew the time of year. At noon the sun would always be high in the sky near the equator, and low in the sky if he was far south or far north. However, working out how far west or east you were was much harder. This made navigation a hazardous business, and ships were in constant danger not merely of getting lost, but of running into land, or rocky reefs, because they did not know their positions. Measuring longitude was a major problem, which remained unsolved beyond the end of the Stuart period (see page 203).

So when the king had an inkling that astronomy might provide the answer, he set up a royal commission to examine the possibility of measuring longitude by using the stars. This would not be possible without a far more precise map of the stars than existed at the time. Charles II then demanded that 'he must have them anew observed, examined, and corrected for the use of his seamen'. This would later involve his new Astronomer Royal, John Flamsteed (see page 177).

The king's interest in and support of scientific endeavour led to the 'Invisible College' being incorporated in July 1662 as the Royal Society – the headquarters of the national scientific establishment, which it remains to this day. In late 1662 Christopher Wren delivered a lecture on what he thought should be the aims of the society: to study meteorology, refractions, and the growth of fruits and grain; plenty, scarcity, and the price of corn; the seasons of fish, fowl, and insects, and epidemic diseases. He said, 'The Royal Society should plant crabstocks for posterity to graft on.' He was a Fellow for the rest of his life, and President from 1680 till 1682.

Not that he was always responsive to the Society; in February 1661 he had been offered the position of Savilian Professor of Astronomy at Oxford. This was the top astronomy position in the country, and he could hardly turn it down, although he had to move back to Oxford, and he increasingly failed to respond to requests from the Royal Society. Indeed, he had such a wide range of scientific interests that he often failed to complete projects, and he rarely published anything; so although everyone thought him brilliant he actually achieved rather little in science. He also made the occasional mistake, which he found embarrassing. In 1658 he wrote a draft paper about the curious appearance of Saturn, first noted by Galileo (see page 170). Wren's telescope was better than Galileo's but still not good enough to discern that the funny 'handles' on Saturn were actually rings.

*Christopher Wren's observation and interpretation of the shape of Saturn.*

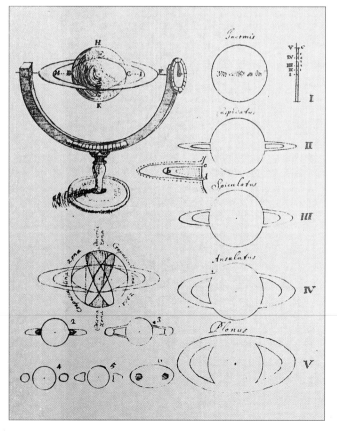

But Wren was immensely popular, and well loved; the Professor of Geometry at Gresham praised him for 'the divine felicity of his genius combined with the sweet humanity of his disposition – formerly, as a boy, a prodigy; now, as a man, a miracle, nay, even something superhuman'. Though his popularity was soured a little by the matter of the globe of the Moon…

Since telescopes had become widely available several people had drawn maps of the Moon, but most were inaccurate, and Wren was sure he could do better. In order to measure precisely the positions of the various mountains and craters he built for himself a modified version of Gascoigne's micrometer (see page 174) which allowed him, while he was looking through his telescope, to move cross-hairs to any position in the field of view and then read off the exact co-ordinates. He then constructed an accurate and beautiful ten-inch globe of the Moon from painted pasteboard, and mounted it on a pedestal of lignum vitae.

When they heard about this globe, the other fellows of the Royal Society suggested that he should bring it to London and present it to the king from the Society, which they thought would be good for its kudos. Wren, however, took a short cut, and without even showing it to the Royal Society presented it directly to the king, with the inscription 'To Charles II, King of Great Britain, France, and Scotland, for the expansion of whose dominions no one Globe can suffice, Christopher Wren dedicates another in this Lunar Sphere'. The king was delighted, but the Fellows of the Royal Society were perhaps a little annoyed.

# Astronomy as a profession

One of the first professional astronomers was John Flamsteed, who was born in 1646, the son of a maltster. A teenage illness prevented him from going to school, and he began to study the stars 'under the discouragement of friends, the want of health, and all other instructors except his better genius'. He observed the solar eclipse of 12 September 1662, and made himself a quadrant. When he observed the eclipse of the sun on 25 October 1668 he commented that 'the tables differed very much from the heavens' – just as Kepler had noted half a century earlier.

Flamsteed's skill brought him into contact with members of the Royal Society and eventually with the king, who was given a barometer, a thermometer, and Flamsteed's rules for using them to forecast the weather. In March 1675 the king appointed Flamsteed the first Astronomer Royal – or rather Astronomical Observer by Royal Warrant – and instructed him 'forthwith to apply himself with the most exact care and diligence to the rectifying the tables of the motions of the heavens, and the places of the fixed stars, so as to find out the much desired longitude of places for the perfecting the art of navigation'.

Flamsteed would clearly need a place from which to carry out his observations, and so the king asked Christopher Wren, wearing his architect's hat, to design a suitable building. Wren chose a position on the hill above Greenwich, and the Greenwich Observatory was built at a cost of £510, which was recovered by the sale of some spoiled gunpowder. Two hundred years later the observatory would become internationally accepted: the meridian, or line of zero longitude, was defined by the cross-hairs of the Astronomer Royal's telescope.

However, when Flamsteed moved in, on 10 July 1676, he found a cold and empty building, devoid of furniture or instruments. He was expected to buy his own equipment, and he continually complained that he could not afford the most precise instruments, although in fact his sextant was the best in the world. During the next twenty years Flamsteed made some 20,000 observations, and although he said they were only preliminary and approximate they were better than any that had been published.

Isaac Newton, working on his theory of gravitation (see page 217), needed accurate astronomical observations to check his calculations, but Flamsteed was unwilling to publish his results until he knew they were as precise as possible. This led to a terrible row, Newton demanding the data and Flamsteed refusing to deliver, until in the end Newton went to Greenwich and literally stole Flamsteed's observations and published them himself in 1712; 400 copies were printed. Flamsteed was so angry that he managed to get hold of 300 copies, and burned them.

*Greenwich Observatory was designed by Christopher Wren and built in 1675 at a cost of £510.*

Although difficult and cantankerous, Flamsteed was a remarkable astronomer. 150 years later, Professor Augustus de Morgan wrote that he 'was in fact Tycho Brahe with a telescope; there was the same capability of adapting instrumental means, the same sense of the inadequacy of existing tables, the same long-continued perseverance in actual observation'. Flamsteed died on the last day of 1719, outliving the Stuarts by five years.

The genial man who was to succeed Flamsteed as Astronomer Royal was Edmond Halley, son of a soap-maker, who left Oxford University without a degree in order to go and study the stars in the southern hemisphere. He realized that if indeed astronomy was going to be useful for sailors, then there would have to be a star map for the south as well as the one Flamsteed was making for the north, since from the southern hemisphere you can see many stars that are not visible from north of the equator, while the pole star, for example, never comes into view.

He chose to stay on the island of St Helena in the south Atlantic, a place so remote that it was where Napoleon was sent in exile after the Battle of Waterloo. This turned out to be an unfortunate choice; the weather was so bad that Halley was able to observe only 341 stars. Nevertheless, he did make the first ever observation of the transit of Mercury, and when he came back to England he was elected a Fellow of the Royal Society, and Flamsteed called him 'the Southern Tycho'.

Halley is remembered, justifiably, for the brilliant prediction he made about a comet. Robert Hooke had explained, in a 1665 lecture, that comets should return in a periodic way, but Halley made a more practical and dramatic prediction. He observed a comet in 1682, and because he was a close friend of Newton, who was in the process of writing up his great book on the laws of motion and the behaviour of heavenly bodies, he was able to use Newton's laws to calculate the comet's path. He must have been excited when he realized that it had been seen before:

> Now many things lead me to believe that the comet of the year 1531, observed by Apian, is the same as that which in the year 1607 was described by Kepler… and which I saw and observed myself in 1682… The identity of these comets is confirmed by the fact that in 1456 a comet was seen, which passed in a retrograde direction between the Earth and the Sun, in nearly the same manner… From its period and path I infer that it was the same comet as that of the years 1531, 1607, and 1682. I may, therefore, with some confidence predict its return in the year 1758.

He was right – just! The comet reappeared on Christmas Day 1758, and since then every seventy-six years; the most recent appearance was in 1985–6. More recent inspection of the records shows that it had also been

*Edmond Halley was the second Astronomer Royal, particularly remembered for his comet.*

observed by the Chinese since 240 BC; it was mentioned in the Nuremberg Chronicles of 684; and it was even displayed in the Bayeux Tapestry after its appearance in 1066. It will next appear in 2061.

What really matters is the prediction. Modern scientific method depends on scientists making observations, trying to deduce general principles, using those principles to make predictions about new phenomena, and then if possible carrying out experiments to test the predictions. A scientist who can make accurate predictions is likely to be doing good science. Halley's prediction of the return of 'his' comet was one of the great predictions of science, and the only sad thing is that he clearly was not going to live long enough to see it come true.

# Calculating worlds

One of the tedious aspects of astronomy is the mathematics necessary to work out the results. Many astronomers have hated doing the mathematics, and none more so than John Napier, the eighth Laird of Merchiston, who lived at Merchiston Castle, south-east of Edinburgh, and now part of Napier University. He was known as the 'Marvellous Merchiston', and locals said that he could tell the future.

He kept a black cockerel that was supposed to be psychic. When some valuables were stolen from the castle, Napier suspected one of his servants, and ordered them to go one by one into a darkened room and stroke the cockerel. He said it would crow when touched by the guilty party. But the cockerel remained silent. Then Napier took them into a lighted room and asked them to hold up their hands. All but one had black hands. The clean-handed servant was accused of theft; he hadn't dared touch the cockerel. Napier had covered the bird in soot.

Napier's favourite pastime was astronomy. However, he found all the calculations tedious, and worked for twenty years on a way to simplify them. In 1614 he produced logarithms, described in his book *Mirifici Logarithmorum Canonis Descriptio*. Using logs, you could multiply by adding and divide by subtracting. Astronomers across Europe thought the technique wonderful, and log tables were rapidly adopted. Most people now over the age of about fifty probably used them at school, before electronic pocket calculators were available.

Logarithms certainly seemed like magic to Henry Briggs, Professor of Geometry at Oxford. When he saw Napier's book he wrote to his friend James Ussher (who later became an archbishop and calculated that the Earth had been created in 4004 BC) to say, 'Napier hath set my head and hands a-work with his new and admirable logarithms. I hope to see him this summer, if it please God, for I never saw a book which pleased me better or made me more wonder.' And indeed he did travel all the way to Merchiston, and met Napier, and the story goes that when the butler showed him in, they sat and looked at one another for fifteen minutes, silent in mutual admiration.

Briggs stayed with Napier for a month in 1615, and another month in 1616, and made an important contribution. Napier's were 'natural logarithms', to base $e$, or 2.718. Briggs worked them all out again to base 10, which made all the calculations easier still.

In his quest to simplify calculations, Napier also invented one of the world's first pocket calculators – some type of abacus was probably first – in the form of a set of rods with numbers on them. Because they were usually made of ivory they were commonly called 'Napier's Bones', and they are remarkably effective for simple multiplication. The rods were like square-section pencils, and to multiply, say, 8 by 17 you took an 8 rod, a

*John Napier, the 'Marvellous Merchiston', who invented logarithms and a pocket calculator*

1 rod and a 7 rod, laid them side by side, and read off the answer from the numbers on the sides. The bones were to be immensely popular for at least a hundred years; Samuel Pepys wrote in his diary in 1667, 'To my chamber whither comes Jonas Moore and tells me the mighty use of Napier's Bones.'

The calculator has evolved ever since then, and taken various forms. The eminent French mathematician Blaise Pascal made one in the middle of the seventeenth century, and it is said that seventy were sold. The first in England seems to have been the curious device produced around 1666 by the eccentric Samuel Morland. He described it as a 'new and most useful instrument for addition and subtraction of pounds, shillings, pence, and farthings; without charging the memory, disturbing the mind, or exposing the operator to any uncertainty; which no method hitherto published, can justly pretend to'. Morland's calculator was a small flat box with dials in the top for farthings, pence, shillings, pounds, tens of pounds, and so on. To add two sums of money you put a pointed implement into each of the dials in turn and moved the dial on by the amount to be added.

Robert Hooke, who thoroughly disapproved of Morland, described this machine as 'very silly', and even Samuel Pepys, who was generally more curious than critical, wrote that it was 'very pretty but not very useful'.

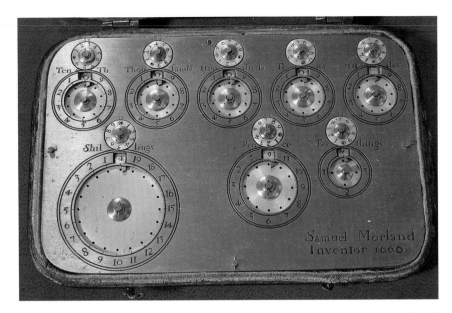

Born in 1625, Morland grew up with the Civil War raging, and joined Cromwell's forces as a spy, since he was the best man in the country at undetectably opening and resealing letters, not to mention copying the contents and forging handwriting. However, when in 1658 he overheard a plot to lure Charles II back to England and kill him, Morland switched allegiance, and rushed over to Holland to warn the king, who rewarded him with a knighthood, a pension, and in 1681 the title of *Magister Mechanicorum*, or Master of Mechanics.

*Samuel Morland's adding machine, described by Robert Hooke as 'very silly'.*

In 1670 Morland unveiled his most spectacular invention, designed for long-distance communication, which he called a *tuba stentoro-phonica*. We should call it a megaphone. Legend has it that Alexander the Great used to summon his troops from 12 miles away with a horn or trumpet, and Morland wanted to develop the idea to relay more complex messages. He built first glass and later brass instruments, with increasing lengths, starting at about 30 inches, and winding up at over 20 feet. He claimed that using the long ones he could carry on a conversation at a distance of three-quarters of a mile, and indeed when it was sent to be tested at Deal Castle in Kent, the Governor reported that speech could be heard at Walmer Castle, more than a mile away. It was also useful for communicating with ships some distance offshore.

So what the Stuarts did for us in the world of science and technology was to take a fundamental step forward in grasping new scientific ideas and then turning them into practical benefits, an approach exemplified in discovering the circulation of the blood. They used the microscope to explore the world of the minuscule; they latched on to the telescope, and developed not only new maps of the heavens but also fundamental ideas about gravity – inventing pocket calculators along the way.

Being the Seventh Chapter

# DESIGNER LIVING

'All accounts of Gallantry, Pleasure, and
Entertainment, shall be under the Article of
White's Chocolate-house; Poetry, under that of
Will's Coffee-house; Learning, under the title of
Grecian; Foreign and Domestic News, you will
have from the St James's Coffee-house...'
LONDON GAZETTE

'Poor, nasty, brutish and short': that was the human condition, according to the political philosopher Thomas Hobbes, who lived until 1679 and so had much experience of life in Stuart times – though as he died at the age of ninety-one, perhaps the 'short' didn't apply to him. Certainly, most people in the seventeenth century led a precarious existence, beset by all manner of disasters. Some miseries were man-made, most catastrophically the Civil War that tore the country apart. Strife could make the constant threat of famine worse, with ravaged land and ruined crops. But there were natural disasters too: drought and flood and, above all, disease.

While the Great Plague of 1665 is a landmark in the history books, in fact the infection was endemic; there were dozens of outbreaks in Tudor times, the worst in 1563, and the Stuarts had had serious outbreaks in 1603 and 1625. The one in 1665 is memorable just for its scale, killing around 100,000 people in London alone. The city's crowded, filthy streets were ideal breeding grounds for the black rats that harboured the fleas that carried the infection. Not that the Stuarts could have known how the disease was transmitted; they just knew it was easy to catch and usually deadly, and probably had something to do with the wrath of God.

*Previous spread: St Bride's Church, built by Christopher Wren. Below: The Great Plague of 1665 caused huge disruption and many people fled the City of London.*

There's a popular belief that the Great Fire of London, which happened the year after the Great Plague, was instrumental in wiping out the disease by the power of purifying flame. In fact, by 1666 the plague had lost much of its virulence, and it struck worst in areas untouched by the fire, though no doubt the conflagration did burn out a few sources of infection. And like plague, fires were a common occurrence in a city of jumbled, mainly wooden, buildings; again, the 1666 disaster was notable for its huge scale.

Two such devastations in the space of just two years would seem to bear out Hobbes's pessimistic observation – yet there were bold and enterprising people resolved to challenge disaster, to struggle furiously to improve their lot, to experience the finer things of life. Today, we might call such aspirations 'designer living': a conscious attempt to improve the material quality of life. And a seminal, enduring example arose from the ashes of the Great Fire of London.

# Building by design

The fire had started in a baker's shop in Pudding Lane, just east of London Bridge, in the early hours of Sunday, 2 September. A strong east wind ensured it spread quickly. Samuel Pepys saw the fire later that morning, and reported 'nobody to my sight endeavouring to quench it, but to remove their goods and leave all to the fire'. By the evening, he reckoned, the front of the fire was a mile long. The next day, John Evelyn wrote, 'they hardly stirred to quench it; so that there was nothing heard or seen but crying out and lamentation, and running about like distracted creatures'. The Royal Exchange burned down that day, and St Paul's Cathedral the next. The Guildhall burned, but the walls survived, as did the City records in the crypt.

Showing a proper regard for priorities, Samuel Pepys buried his papers, his wine and a Parmesan cheese, and then saved the Navy Office (where he worked) by getting all the surrounding buildings blown up with gunpowder, and so creating a fire break. The Tower of London was saved in the same way. By Thursday the fire was under control, partly as a result of the efforts of the king and his brother the Duke of York, and partly because the wind had died down. So the fire burned for just four days. Only nine deaths were recorded, but the damage to property was unprecedented: over 13,000 houses and many public buildings were utterly destroyed, including eighty-seven churches. The total loss was estimated at £10 million.

People were gradually rehoused, but the fabric of the city had to be rebuilt. Grand plans were quickly put forward by John Evelyn, Robert Hooke, Christopher Wren and others, but money was limited, and there were complications about the ownership of land and previously existing buildings. The task of organizing the rebuilding was given to a six-man committee, the Commissioners for Rebuilding, which included Hooke and Wren, both of whom were distinguished members of the Royal Society (see page 162).

Wren was a Professor of Astronomy, and had already displayed an extraordinary aptitude for science and technology, but he had also designed two fine buildings: a new chapel for Pembroke College Cambridge, and the Sheldonian Theatre in Oxford, where degree ceremonies are still held. He was only an amateur architect, but he was an extraordinarily fine draughtsman; as Robert Hooke wrote in the Preface of his great book

'...they hardly stirred to quench it; so that there was nothing heard or seen but crying out and lamentation, and running about like distracted creatures...'
JOHN EVELYN ON THE GREAT FIRE

*Micrographia* (see page 165). 'I must affirm that, since the time of Archimedes, there scarce ever met in one man, in so great a perfection, such a Mechanical Hand, and so Philosophical a Mind.'

Wren was undoubtedly influenced by the great architect Inigo Jones, who was born in Smithfield, on the edge of London, in 1573. His father was also called by the unusual name Inigo, which was probably Welsh. Young Inigo became a 'picture-maker' or professional artist. He spent some years in Venice around 1600, wrote notes on Stonehenge which were turned into a book by his pupil John Webb, and then from 1605 he began designing costumes for masques in England.

Masques were short intense theatrical productions, popular with Queen Anne and with others of the court. They were incredibly lavish and expensive, the text often written by Ben Jonson, while Jones frequently designed the costumes and sets. He seems first to have turned his artistic hand to architecture around 1608. Before then building had generally been supervised by the builders, the stonemasons and the carpenters, and the idea of an artist drawing up an overall design seems to have been rather novel. His early designs show the influence of the famous Italian Andrea Palladio, who wrote a treatise *Quattro Libri dell'Architettura* in 1570. Jones designed a wonderful romantic palace for the masque *Oberon* in 1611, for young Prince Henry (who was to die in 1612, precipitating his brother Charles to the throne); then in 1613–14 he had the chance to go on a grand trip to Italy, and took the trouble to visit many of the buildings described by Palladio, and also absorbed ideas from other Italian masters.

When he came home, and James I made him Surveyor of the King's Works, with almost unlimited funds, Inigo Jones took wings. He started modestly with a few small buildings in Newmarket, where the king enjoyed hunting, and then built a new house for the queen at Greenwich; this has ever since been called The Queen's House. He built the New Exchange in the Strand, the Queen's Chapel in St James's Palace, and the banqueting house in Whitehall – unfortunately remembered as the building from which Charles I walked on to his scaffold, and sadly the front was completely rebuilt by Sir John Soane 200 years later. In ten years Jones designed some sixteen buildings for the royal family and their close friends.

Some of these had almost comic ends. For George Villiers, Duke of Buckingham, he build a large and elegant mansion, York House, next to what is now Charing Cross railway station. After the great fire Buckingham agreed to sell the house to a property developer, Nicholas Barbon, on condition that his name was preserved. So if you go there today you will find George Street, Villiers Street, Duke Street, Buckingham Street and even Of Alley, renamed York Place.

English architecture lagged well behind the classical ideas that permeated the Continent, emanating especially from Italy, mainly because Henry VIII

had driven such a chasm between reforming England and Catholic Europe. Inigo Jones, picture-maker, costume designer and Surveyor of the King's Works, was the first man to bring classical Italian architecture to England. One important innovation he brought back with him was a new roof structure called the king post truss. Tudor roofs had relied on simple struts from the main crossbeams to hold up the rafters and the roof, but this meant that the beams had to take all the strain, and often large roofs collapsed under their own weight.

The king post truss was simple and clever. The central post was held up by the rafters pushing up under its 'shoulders', and the struts from the rafters rested on its 'knees', so that the post was in tension, and supported the weight. This took the weight off the crossbeam, and made the truss much stronger, so that roofs could be made much larger.

Christopher Wren was to take this idea and develop it into the queen post truss, with two upright posts and a flat roof above them. He also supported the crossbeams by tying them to the posts with hefty iron straps. Then he was able to suspend a heavy ceiling below the crossbeams, and the

*Inigo Jones brought classical Italian architecture to England.*

CHRISTOPHER WREN Kᵗ
PRESIDENT OF THE ROYAL
SOCIETY

*Admired by everyone, Sir Christopher Wren was a mathematician, scientist and architect, and a friend of Robert Hooke and Edmond Halley.*

Painted Hall in Greenwich is a lasting memorial to his skill. The Guildhall in Windsor has another superb suspended ceiling, but the story goes that the contractors did not trust Wren's judgement, and insisted that he insert pillars to hold up the ceiling, which they were sure would collapse. He was so angry that he deliberately made the pillars an inch too short, and there is still an inch gap between the pillars and the ceiling.

Despite Inigo Jones's achievements, he is almost forgotten today. This is partly because only seven buildings survive of the forty-nine he designed, and partly because he was overshadowed by Wren, who got his chance for preferment as a result of a bedroom farce. The fifty-five-year-old Surveyor General Sir John Denham had married a beautiful eighteen-year-old called Margaret Brooke, who immediately hopped into bed with the Duke of York, and their affair was so public that Denham went slightly mad, marched into the king's chamber and announced that he was the Holy Ghost. Then his pretty young wife died suddenly; most people were convinced she had been murdered either by the Duchess of York or by her husband – and Denham was so humiliated that he died in March 1669. The king needed a young, energetic and capable man to take over the rebuilding of London and, remembering the fine work he had seen in the past, he sent for Christopher Wren, and invited him to be Surveyor General of the King's Works. Wren was just thirty-seven at the time, and his appointment was not received with general enthusiasm, especially by those who had been passed over; but it turned out a great success.

For a year or two he still had time to go to the Royal Society, but then a fat tax was slapped on coal in order to raise a great deal of money, and rebuilding could really gather pace. Over the next thirty years he and Robert Hooke worked closely together. These two were joined by Nicholas Hawksmoor, who became Wren's assistant in 1679, and together they built not only St Paul's Cathedral but fifty-one other churches, whose spires graced the London skyline. Wren was particularly careful with the tall delicate towers, which he often left until the rest of the church was finished, to make sure he got them just right.

St Paul's started as a nightmare. Wren's first design was considered not grand enough. His second delighted the king, who knighted him – but the clergy disapproved. Finally in 1675 he produced a third design; the king approved, and issued orders that Wren was to go ahead; so naturally the clergy liked that one. Wren had been careful to ensure that his royal warrant gave him 'the liberty in the prosecution of his work to make variations, rather ornamental than essential, as from time to time he should see proper'. This gave him carte blanche, and he proceeded to put up the building he wanted. Whenever anyone raised a query, he said it was just a minor variation.

St Paul's was finally completed in 1708, and Wren died in 1723, outliving the Stuart dynasty. St Paul's Cathedral was a large part of Wren's life; it inspired the clerihew (from E.C. Bentley's *Biography for Beginners*):

> Sir Christopher Wren
> Said, 'I am going to dine with some men.
> If anybody calls,
> Say I am designing St Paul's.'

The magnificent building also became his monument in death: he was buried there, and a Latin inscription reads *Lector Si Monumentum Requiris Circumspice*: 'Reader, if you want a monument, look around you.'

'I must affirm that, since the time of Archimedes, there scarce ever met in one man, in so great a perfection, such a Mechanical Hand, and so Philosophical a Mind.'

ROBERT HOOKE ON CHRISTOPHER WREN

Wren did not rebuild only religious houses; other projects were in a decidedly secular world, including the Theatre Royal, Drury Lane, which was probably where King Charles met one of his favourite mistresses, Nell Gwyn, who herself represented a revolution on the English stage.

# Acting by design

In Shakespeare's time, women were forbidden to act on the stage; women's parts were played by boys, and frequently he had boys dressed as girls who then dressed as boys to complicate the plot. The Stuart theatre saw the first actresses on the public boards.

During the Commonwealth, Puritans held theatres, like many other pleasurable pursuits, to be sinful; in 1642 Parliament issued an ordinance closing all theatres and forbidding public performance of plays. They remained closed for at least fifteen years, although things improved slightly in the late 1650s. However, it was not until July 1660 that theatres were officially allowed to open once more, thanks to Charles II, who gave William Davenant and Thomas Killigrew the right to organize two companies of actors in London, so that they essentially acquired a monopoly over the theatre. According to rumour, Davenant may have been an illegitimate son of William Shakespeare, who used to pass through Oxford on his way to and from his home at Stratford, and more than once shared a bed with William's beautiful mother Jennet.

In 1663 Davenant and Killigrew acquired patents that declared:

> *Forasmuch as many plays formerly acted do contain several profane obscene and scurrilous passages and the women's parts therein have been acted by men in the habit of women... henceforth no new play shall be acted... containing any passage offensive to piety and good manners... And we do likewise permit and give leave that all the women's parts... may be performed by women so long as these recreations... be esteemed not only harmless delights but useful and instructive representations of human life...*

One of the first women to appear on the stage was Margaret Hughes, who performed as Desdemona in 1663 at the Theatre Royal. Around 1669 she was carried off to be mistress of Prince Rupert, and disappeared from the stage until 1676. Another early actress was Mary Betterton, and then came the three Elizabeths: Boutell, Cox and Barry. They were all popular both

with the critics and with the audiences, but the best-known of all was Nell Gwyn, who started working life by following her mother and her sister into prostitution, but became an 'orange girl' in the theatre; to the male customers she sold oranges and other favours, including introductions to racy women in the audience. As John Dryden wrote:

*Nell Gwyn was a popular actress and favourite mistress of Charles II, whose last words were 'Let not poor Nelly starve'.*

> *The play-house is their place of traffic, where*
> *Nightly they sit to sell their rotten ware…*
> *For while he nibbles at her amorous trap*
> *She gets the money, but he gets the clap.*
> *Entrenched in vizor masks they giggling sit*

*And throw designing looks about the pit,*
*Neglecting wholly what the actors say.*
*'Tis their least business there to see the play.*

'Pretty, witty Nell' was given her first chance on the stage about 1665, and was lucky to be taken up by Dryden, who in due course wrote several parts specially for her, to take advantage of her electric presence and her native wit. She became a talented and much loved actress. Pepys took his wife to see her perform one night, and they both found her such a 'mighty pretty soul' that neither he nor his wife could resist kissing her. She met King Charles some time in 1669; according to one story they were watching a play in adjacent boxes. According to another, she was acting, and recited a monologue wearing a hat the size of a large coach-wheel, which suited her to perfection. In both stories Charles was induced to invite her immediately to supper, and thence to bed.

While Charles was a pleasure-loving monarch, and pretty tolerant on the whole, like many another ruler he feared plots and sedition. By the time he came to the throne, there had been a development in the nation's social life that he thought represented a threat to security: coffee-houses. To those of us today who happily frequent Starbucks and the like, it may seem bizarre that a place of innocent refreshment could be seen as a hotbed of revolution.

*Opposite: The coffee-houses of London were hotbeds of gossip, intellectual discussion and political debate.*

'In Covent Garden tonight I stopped at the great Coffee-house, where I never was before... it will be good coming thither; for there I perceive is very witty and pleasant discourse.'

SAMUEL PEPYS

## Coffee-houses by design

Coffee was one of the fancy new comestibles introduced to England in Stuart times. The first coffee-house was opened in Oxford in 1650; two years later they began to appear in London, and then elsewhere in the country; by the 1660s they were pretty well established. Customers generally had to pay a penny for a cup, and the coffee-house was sometimes called the 'penny university', reflecting the intellectual stimulation visitors could expect. The habitués were almost entirely male, and fairly well-off. Inevitably, men of various professions and political persuasions came together at particular haunts. Robert Hooke loved to frequent coffee-houses with Christopher Wren and Edmund Halley. Steele's *London Gazette* mocked the tendency: 'All accounts of Gallantry, Pleasure, and Entertainment, shall be under the Article of *White's Chocolate-house*; Poetry, under that of *Will's Coffee-house*; Learning, under the title of *Grecian*; Foreign and Domestic News, you will have from the *St James's Coffee-house*...'

Coffee-houses served as libraries and debating chambers; they provided periodicals – the *Tatler* and the *Spectator* both started in coffee-houses – the latest polemical best-sellers, and customers to pass passionate judgement on them. Jonathan Swift wrote of the 'Committees of Senators

who are silent in the House, and loud in the Coffee-house, where they nightly adjourn to chew the cud of politics, and are encompassed with a ring of disciples who lie in wait to catch up their droppings'. Members of the Royal Society frequented them, and discussed intellectual developments alongside social scandal and religious upheaval. Pepys spent many an evening in them; in February 1664 he recorded, 'In Covent Garden tonight I stopped at the great Coffee-house, where I never was before; where Dryden the poet (I knew at Cambridge), and all the wits of the town, and Harris the player, and Mr Hoole of our College [i.e. Gresham]. And had I time then, or could at other times, it will be good coming thither; for there I perceive is very witty and pleasant discourse.'

Some coffee-shop customers were content to sit all day over a single cup, but those who wanted quick service could pay for it; according to legend there was a box on display labelled TIP, for To Insure Promptness. The impatient put money in the box, and the waiter came quickly. That is one version of why today we leave a tip.

In the smoky corners of the coffee-houses, not only were tea and coffee and hot chocolate consumed in huge and sugary quantities, but rumours were spread, opinions aired, deals struck, friendships made and broken. 'Here they treat of Matters of State, the Interests of Princes, and the Honour of Husbands, &c. In a Word, 'tis here the *English* discourse freely of every Thing, and where they may be known in a little Time.' Women were up in arms, partly no doubt because they were effectively excluded. The Women's Petition Against Coffee accused the coffee beans of making men sterile and impotent – but it met with a spirited counterblast in the shape of The Men's Answer to the Women's Petition. Charles and his government were worried about all the potentially seditious gossip, which might easily lead to plotting. The king used the women's petition as an excuse to order all the coffee-shops to be closed down. However, this raised such a storm of protest that he was forced to relent, and they stayed open.

The coffee-houses bred one particular enterprise: the insurance market. In these establishments shipping merchants would hear about loads to be bid for, or the misfortune of a fellow trader attacked by a pirate at sea. The cafés and the ships were linked directly by tea and coffee and sugar but also, more indirectly, by financiers. The money men had realized that the

THE
WOMEN'S
PETITION
AGAINST
COFFEE.
REPRESENTING
TO
PUBLICK CONSIDERATION
THE
Grand INCONVENIENCIES accruing to their SEX from the Excessive Use of that Drying, Enfeebling LIQUOR.
Presented to the Right Honorable the Keepers of the Liberty of VENUS.

By a Well-willer

London, Printed 1674.

huge risks inherent in shipping could be turned to profit, could be calculated and reduced to figures. Fire insurance was born in the 1660s, but did not prosper as the slightly more mature marine insurance market did. One of the biggest names in the business, Edward Lloyd, gave the 200 or so faceless underwriters more of an identity when in 1688 he founded a coffee-house where they could meet shipowners, captains and merchants, who could be encouraged to sell their souls with complimentary cups of coffee; so was born Lloyd's of London. Even today the uniformed attendants at Lloyd's are called waiters. (See also *Henry Winstanley and the Eddystone Lighthouse* by Hart-Davis and Troscianko, 2002.)

Some people have even suggested that England's Industrial Revolution was seeded in the coffee-house. All manner of men met and mingled here, including scientists, inventors and investor–patrons: a fertile ground for new developments and enterprise – unlike, say, the more stuffy, stratified society of France, where wealthy aristocrats might not deign to socialize with humbler merchants and artisans. As we have seen, this was an age where scientific thought and invention were gathering pace. One specialized area of interest was glass, something we take for granted now – indeed, some modern buildings seem to be made of little else.

# Glass by design

Most English glass-makers in the early seventeenth century were essentially using Roman techniques and chemistry, and the glass was ripe for improvement. The Royal Society included several members of the Worshipful Company of Glass Sellers, who were keen to use their scientific skills and contacts to try and find a way of making glass as beautiful as the clear or 'cristallo' glass from Venice. They commissioned George Ravenscroft to investigate. Glass is made mainly of silica, which is chemically the same as sand, with various additives. Ravenscroft started by crushing flints to make his silica, and the resulting glass was called flint glass, but it had an unfortunate tendency to 'crizzle', forming minute cracks.

He managed to overcome this crizzling by adding lead oxide and a little potash to the silica, and when he reached about 30 per cent lead oxide he got a wonderful soft, heavy glass with a high refractive index, which meant that it looked superb when it was cut – clear, brilliant and sparkling. He patented his process in the mid 1670s, and this glass has ever since been called crystal, lead crystal, or flint glass. This was a classic example of what Francis Bacon had written about fifty years before (see page 160): the use of scientific experiments to bring about benefits for mankind, a clever application of the growing science of chemistry. Another example was the production of alum; it's worth spending some time with this substance, as it provides a fine example of the early chemical industry in England – and the natives' ingenuity.

'Here they treat of Matters of State, the Interests of Princes, and the Honour of Husbands, &c. In a Word, 'tis here the English discourse freely of every Thing, and where they may be known in a little Time.'
JOHN EVELYN

# Dyeing by design

An unexpected result of Henry VIII's split with Rome was that the Vatican cut off supplies of alum; this may not sound serious but it did present a problem, because without alum the English could not dye cloth – and so throughout his later life Henry VIII had dull and colourless socks.

Alum is a mordant, a chemical that binds strongly to the fibres of wool, cotton and linen, and also to molecules of dye. If you try to dye these materials without a mordant the colours are never bright, and the dye is not fast; it easily washes out. So for hundreds of years dyers had heated their cloth in a solution of alum before treating it with dye. In the 1450s the Pope's envoy Giovanni de Castro discovered an extensive alum supply in the Tolfa Hills, close to Rome – 'I have… found seven hills so stocked with alum as to be nigh sufficient for seven worlds' – and the Vatican took control.

By then, alum had been in use for at least 3,000 years; it is mentioned in the Ebers papyrus of about 1500 BC, and was used not only in dyeing but in a variety of other applications; the Roman writer Pliny said a good deal about it, although he was rather vague and inaccurate. A medieval manuscript mentions alum for treating leather to make it both tougher and more supple, and for writing in gold. Treatment with alum improves parchment and paper, and it is even possible that some early alchemists thought of alum as the philosopher's stone that could turn base metals into gold.

With the possible exception of gold itself, alum was the only material that the ancients were really able to purify. In general they had no clear idea of what purity meant, and until Robert Boyle coined the modern definition of an element in 1661 the concept was difficult.

Alum dissolves a little in cold water, but much more easily in hot water. Dissolve as much as you can in hot water and then let the solution cool, and the alum will crystallize: large, beautiful octahedral crystals grow as if by magic in the flask, and these crystals will be much purer than the original material. This process can be repeated, and the recrystallized alum should be completely pure. Crude alum is often contaminated with salts of iron, and because these are always coloured, the mixture usually has some colour. Pure alum, however, is colourless; the crystals are clear, and this absence of colour is to some extent an indication of purity.

When the supply of alum from Rome was cut off, dyeing in Britain became impossible. The only way to get cloth coloured was to send it to dyers in Flanders (Belgium), who charged a lot of money but did not do a very good job. So the search was on from the 1540s to find another source of alum. It has been said that one reason why Henry VIII married Anne of Cleves was to get his hands on the alum supplies in Flanders, though this was not a successful enterprise. He also started looking for sources closer at hand. At least six major attempts failed, in such places as Ireland, the Isle

of Wight (Alum Bay), and Dorset (Alum Chine). Success finally came at the end of the sixteenth century, through the Chaloner family.

The second Sir Thomas Chaloner travelled to Germany, where alum was made, and in 1596–7 to the Tolfa Hills, and seems to have returned with the secrets of the process. One story, told by the writer John Aubrey, is that while he was riding in Yorkshire, on a common, 'he tooke notice of the soyle and herbage and tasted the water, and found it to be like that where he had seen the allum workes in Germanie. Whereupon he got a patent from the King...' However, Singer in his book on the alum industry says that Aubrey's account is 'a garbled memory of an inaccurate account of Pius II's embroidery of de Castro's vision of his discovery at Tolfa' – so perhaps we should not take it too seriously!

Another story is that Chaloner smuggled Italian workmen from Tolfa in a barrel and brought them back to England. He may also have read *De Re Metallica*, published in 1556 (see also page 62). It gives three separate methods of making alum then in use in Flanders and Germany, and Chaloner was almost certainly familiar with them. At any rate in the early 1600s Chaloner acquired a patent from Queen Elizabeth to be the sole producer of alum.

*Hanks of wool dyed and then rinsed. The one on the right was pretreated with alum.*

*Quarry in the cliffs at Ravenscar, with the remains of the Peak Alum Works in the foreground.*

Thirty years were to pass before England actually became self-sufficient in alum, but in due course an immense industry grew up along the bleak coastline of north-east Yorkshire, either side of Whitby. The cliffs were attacked along a 30-mile front from Lofthouse and Boulby to Ravenscar; even by 1608 some £20,000 had been invested in the process. The annual production of alum grew from about 700 tons in 1616 to 1,800 tons in 1635, and thousands of people were employed for about 250 years in this, the first chemical industry in England. One of the most remarkable aspects was that chemistry was overtaking alchemy: these alum makers knew what they were doing, even though they had no modern notions of elements, compounds or chemical reactions. The ingredients for making alum were grey shale from the cliffs, water, and either toasted seaweed or stale human urine.

On the cliffs, labourers used pickaxes to remove the top layers of rock and clay, then they hacked out the shale and trundled it down in wheelbarrows to the base of the cliff where they piled it on huge bonfires – alternate layers of brushwood and shale 10 yards wide, 50 yards long, and perhaps 50 feet high. The fire was lit and kept burning slowly for about nine months, the outside being damped down and sometimes covered with clay to prevent it from burning too quickly.

The alum shale is a mixture of clay, iron pyrites (fool's gold), and other minerals. The vital ingredients are the alumino-silicates in the clay and the sulphur in the pyrites. The slow roasting was necessary to bring about a series of complex chemical reactions. First the iron pyrites is oxidized by the oxygen in the air to make sulphuric acid (and some sulphur dioxide). Secondly, the sulphuric acid reacts with the alumino-silicate to make aluminium sulphate. Controlling the rate of burning was vital; if the fire got too hot all the sulphur might be converted to sulphur dioxide gas and driven off, so that no sulphuric acid was formed.

After nine months of roasting, the remains of the fire were tipped into great tanks and thoroughly washed with water. This leaves behind most of the silicate and other unwanted material, and produces a solution of crude aluminium sulphate in water. Aluminium sulphate is a perfectly good mordant for dyeing, but it is difficult to purify; getting rid of contaminating iron salts is tricky. So the aluminium sulphate was converted into alum, by the addition of either toasted seaweed or stale human urine.

Alum is actually a double sulphate of aluminium and either potassium or ammonia. Those Elizabethans who started the alum production would have got a little potassium into the mix from the wood they burned on the fire, but they needed more potassium, and they imported it from as far away as Orkney and Ireland in the form of toasted kelp. The other option was ammonia, which is formed as urine decomposes. To begin with, they collected urine locally, from barrels at every farm in the district. However, as production increased, demand outstripped supply, and they had to ship urine in from Newcastle, from Hull, and even from London. It is said that the first public urinals in Hull were built in order to collect urine for the alum production up the coast. In London, buckets were placed on street corners and men invited to contribute. Once a week a horse carrying two barrels would go round to collect the offerings, and the barrels were shipped up the North Sea, to be landed with care on the rocky coast. Not an enviable task; as the urine decomposed the smell must have got worse and worse. Daniel Colwall wrote in 1678 that 'the best urine comes from labouring people who drink little strong drink'.

The stale urine or the seaweed ash was added to the solution of aluminium sulphate to make the 'liquor', which was then allowed to flow down channels into the boiling house, where in great lead pans resting on iron bars it was heated by a coal fire underneath. And now the conditions were set for the elegant process of purification, so necessary if the alum was to be useful.

The alum exists in solution as separate entities – sulphate ions, potassium or ammonia, and aluminium ions, all surrounded by water molecules – but as the water is evaporated by the heat, so the solution becomes stronger and stronger. Eventually it reaches the ideal concentration, at which, when allowed to cool, the alum will crystallize pure, and in particular not

'...the best urine comes from labouring people who drink little strong drink...'

DANIEL COLWALL

*Crystals of alum.*

contaminated by the iron salts that are always present. If the evaporation is overdone, then the iron salts will also crystallize, and contaminate the alum crystals.

According to legend, the alum makers had a closely guarded secret, a foolproof way of knowing just when to stop the evaporation and begin the process of cooling: they put a fresh chicken's egg in the tank. To begin with it would sink to the bottom, as it does in a saucepan of water, but during the evaporation the liquor becomes more and more concentrated and more and more dense, until there comes a time when it is more dense than the egg, at which point the egg floats to the surface. This stage just happens to be the ideal concentration for crystallization of pure alum: the floating egg is the signal for the workers to start the cooling stage. The liquor was left to cool for four days to give up its precious cargo of crystals.

This was the curious mixture of alchemical dreams and proto-chemical ideas that kept the first chemical industry going until the middle of the nineteenth century. The scale of the enterprise was extraordinary. Every year thousands of tons of alum were produced. For each ton of alum they had to dig some 15 tons of shale from the cliff, and for each ton of shale they had to remove 3 tons of rock on top of it – all by pick and barrow. What is more, for each ton of shale they had to bring in 6 tons of coal, and 3 tons of dry seaweed or 2 tons of urine (which is about as much as three or four people produce in a year).

The town of Sandsend was built entirely for the workers, and Whitby became important as a port because of the alum being shipped out. In the nineteenth century, however, Victorian ingenuity produced not only a cheap way of making sulphuric acid but also the first synthetic dyes, and suddenly alum was a product whose time had gone, and the whole industry died away.

What is amazing about the alum story is how all that chemistry was discovered, at least 150 years before chemistry was officially invented. None of the processes was particularly complicated, but how on earth did anyone discover that they had to roast the shale for so long and then wash out the ash? How did they discover what to add to turn the product into alum? Many chemical reactions have been discovered by accident – including the making of the first synthetic dye – but it beggars belief that anyone would hack a bit of grey rock out of a cliff, roast it for nine months, wash it, and then add stale urine, just to see what happened…

Other inventions came about in a much more straightforward, predictable fashion, notably one towards the end of the Stuart period – which again was in the spirit of scientific aspiration to improve the quality of life; this time, life at sea.

# Lighthouse by design

Throughout the Tudor and Stuart centuries, more and more ships were taking to the sea. Trade was stretching further and further around the world, and steadily increasing in volume, and the Navy was also growing, partly to defend the merchant ships and the home islands, but equally to support wars of aggression.

For all these ships, accurate navigation was important, and the problem of longitude became ever more pressing. Finding out how far north or south you were was fairly simple (see page 175) but, if you did not know how far east or west you were, you could easily crash into an island in the dark. On 22 October 1707, when the fleet was returning from Portugal under the command of Admiral of the Fleet Sir Clowdisley Shovell, they came out of the Bay of Biscay at night under full sail, and as a result of a simple navigational mistake careered straight into the Scilly Islands. Four ships were lost; 2,000 men drowned; and Shovell himself crawled ashore in a bad way, only to be murdered on the beach in Porthellick Cove by a woman who stole the emerald ring off his finger.

This tragedy triggered a new drive to conquer navigation, and Queen Anne announced a prize of £20,000 for anyone who could solve the

*Sir Clowdisley Shovell, Admiral of the Fleet, who lost four ships, and his life, on the Scilly Isles.*

problem of longitude at sea. This produced many weird and wonderful proposals, involving astronomy, cannons and wounded dogs, but the man who finally won the prize fifty years later was John Harrison, a village carpenter and clock-maker, who designed and built a successful marine chronometer.

Another way to reduce the perils facing sailors was to build lighthouses. These aids to shipping had been built since ancient times, most famously the Pharos of Alexandria. More recently, in England, Henry VIII had set up Trinity House as the authority charged with organizing lighthouses around the coast (see page 55), but by the end of the seventeenth century they had achieved rather little. One big problem was that the wreckers of Devon and Cornwall made their living by plundering shipwrecks, and they did not want any lighthouses to get in the way of their trade. They had been known to lure ships on to savage rocks by rigging false lights, masquerading as harbour entrances, or even hanging lights from horses' tails to look like ships swaying at anchor.

One particularly vicious reef was the Eddystone Rocks, 14 miles south-south-west of Plymouth and right in the path of ships heading for the port. Since records began, a ship has been wrecked for every hundred yards of Devon coastline, but no one knows how many have gone down on Eddystone, for neither wreckage nor survivors are often washed ashore from so far out to sea.

In 1695 two ships went down on the Eddystone, the *Snowdrop* in the summer and the *Constant* on Christmas Eve. Two battered survivors from the *Constant* made their way to London and reported the loss to the ship's owner, Henry Winstanley, engraver, showman, and friend of the previous king but one, Charles II. Born in Saffron Walden in Essex in 1644, Winstanley had been employed as porter at the vast sprawling stately home of Audley End when Charles II bought the house in 1669; the king appointed him Clerk of Works. In the neighbouring village of Littlebury Winstanley built himself a house full of practical jokes – a chair that imprisoned your arms, another that catapulted you backwards through the french windows and suspended you over the moat, and a pair of slippers that when kicked released a ghost from the floorboards. He charged visitors a shilling to see 'Winstanley's Wonders' and he became quite rich.

He built a spectacular water theatre at the Hyde Park end of Piccadilly in London, and became richer still. He bought five ships, and when he heard of the fate of the *Constant* he rode at once to Plymouth to find out what had happened. The fishermen and the mayor told him that the king (by now William of Orange) had brought his entire fleet into Plymouth, and was only too well aware of the danger posed by the Eddystone; indeed he had strongly suggested that a lighthouse should be built on the reef – but who was going to build it? 'I will,' said Winstanley.

No one had ever before built a lighthouse on a rock in the open sea and, if he had had any idea of the difficulties ahead, Winstanley would never have volunteered. However, he was vain, and irrepressibly confident, and determined both to make his mark on the world and to impress the king, and he set to work with extraordinary determination.

He started in the summer of 1696. He would set off with his men from Plymouth at high tide, in order to get out of Plymouth Sound, row south, and hope to pick up the main ebb tide flowing out of the English Channel. Then, on a reasonable day, they could row out to the reef in six hours. Usually when they got there the sea was too rough to allow them to land, for Eddystone creates its own swirls and eddies and currents in the tides, and even the slightest Atlantic swell can make landing from a boat impossible. So they had to row back again – twelve hours' rowing for nothing.

On the occasional good day they could land for a few hours at low tide, and go to work on the rock. There was only one rock big enough to be the base for a lighthouse; it became known as the house rock. It is about 10 metres wide, slopes up at about 30 degrees, and stands about 3 metres above the water at high tide. On the lee side of this rock there is deep water, and at low tide a small ledge that can act as a tiny landing stage. The first task was to attach something firmly to the rock. The men took pickaxes and hacked furiously, but the red rock is tougher than concrete, and the mightiest blow from a pickaxe chipped off only a tiny sliver of rock. The men were continually wet from spray; the pickaxe handles were slippery, and any tool dropped over the edge would sink without trace.

In the whole of that first summer all they managed to do was dig twelve holes in the rock. Then they made a fire and melted lead, and stood in each hole a great iron bar as high as a man and as thick as his arm. They poured molten lead around the irons to fasten them into the rock, and then they had to leave before the autumn weather closed in. Only twelve holes! The doubters said it was impossible; no one could build a lighthouse on a rock in the open sea.

The following summer Winstanley and his workers began the awesome task of ferrying great blocks of stone out to the house rock, winching them up, and cementing them between the iron bars. They had been doing this for a few weeks when a French privateer sailed up, and armed men swarmed on to the rock. They stripped the men naked and turned them loose in an open boat; they threw into the sea all the laboriously built-up stones; and they hauled Winstanley off to France and clapped him in jail.

The Admiralty sent a stiff note to Louis XIV protesting that this behaviour was not acceptable, and the Sun King graciously set Winstanley free and sent him home, saying, 'We are at war with England, not with humanity.' He had the sense to realize that the lighthouse would be as useful for the French sailors as for the English. Winstanley went back to work, but weeks had been wasted, and the critics crowed that the lighthouse would never be built.

That second summer he managed to build a stone base 16 feet across and 12 feet high before the autumn gales stopped work for the year. In 1698 he built a magnificent octagonal wooden pagoda on top of his stone base, with sleeping accommodation and a glass-walled 'lanthorn' at the top. In the evening of 14 November he climbed up and lit sixty candles there. The candles may have smoked and flickered in the draught; the light may have been dim; but in Plymouth there was great rejoicing, because for the first time ever the fishermen knew in the dark where the Eddystone was. The only person who did not join in the celebration was Henry Winstanley; the weather was so bad that it was five weeks before he and his crew could get back to shore, by which time they had run out of food.

That enforced stay in the lighthouse gave him a good idea of how bad the winter storms could be. Some waves broke over the top of the lanthorn, even though it was 80 feet above the sea. The structure rocked so much

'My crowning wish is to be in my lighthouse for the greatest storm that ever blew.'

HENRY WINSTANLEY

EDY-STONE LIGHT-HOUSE BLOWNDOWN IN
the Project HEN: WINSTANLY Gen:
1703.
Lost There at y: Time,

*Henry Winstanley's Eddystone Lighthouse.*

in the wind that crockery fell off the table, and some of the men became seasick. Accordingly, the following year he rebuilt the whole thing – a higher stone base and a taller tower, with a total height of 100 feet.

Still the doubters said that the lighthouse would not survive a really bad storm. However, for five years the light did its warning duty, and no ship was wrecked on Eddystone. Then came the Great Storm of 1703, which was probably the most severe in recorded history.

In the middle of November began two weeks of relentless gales from the west. All the ships from the Atlantic arrived early and took shelter where they could along the south coast. All the ships that had been preparing to leave remained in port, unable to set sail into the teeth of the westerly gale. So the seas were swept clean of ships, but every port, harbour and estuary was crammed with vessels, moored to jetties, or to one another, or just anchored in open water.

On Thursday 25 November the wind died down, and Winstanley decided to go out to the lighthouse to make sure everything was shipshape and to supervise repairs for the winter. As he stepped into his boat at the Barbican in Plymouth the fishermen warned him that the storm was not over, and advised him not to go. But Winstanley was fed up with the critics, and with fishermen telling him what he could and could not do, and he boasted, 'My crowning wish is to be in my lighthouse for the greatest storm that ever blew.'

His wish was cruelly granted.

He sailed out to the lighthouse, his boat went away, and the next night the wind got up again. The young journalist Daniel Defoe, who later became famous for his book *Robinson Crusoe*, recorded the damage done in his book *The Storm*. He rode out into Kent and counted 17,000 trees uprooted before he stopped counting. He gathered reports from around the country of lead from church roofs being rolled up like tissue paper and blown away, of men and animals being lifted off their feet, of 800 houses being totally destroyed, and of all those ships being blown on to the rocks, on to one another, and even on to the land. Eight thousand sailors drowned that night, within yards of the shore.

Henry Winstanley was in his lighthouse, which was seen at midnight to be showing its light as usual. But when the sun came up on Saturday morning there was no sign that the lighthouse had ever existed, apart from a few pieces of twisted iron, sprouting out of the rock…

So what the Stuarts did for us in these diverse areas was to carry on from the Tudor idea of 'the good life': to try and rise above the tide of man-made and natural misery. On the theatrical front they introduced greater naturalism to the English stage by allowing women to act in public. They enhanced building design – and indeed the layout of the city of London itself – and improved the glass industry; their coffee-houses were a forum for argument and endeavour, and gave rise to the insurance industry that is so much a feature of modern life; they started the chemical industry by producing alum, and they strived to make the seas safer with an innovatory lighthouse.

Being the Eighth Chapter

# THE APPLIANCE OF SCIENCE

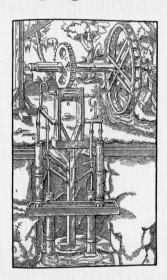

'I do not know what I may appear to the world, but to myself I seem to have been only like a boy playing on the sea-shore, and diverting myself in now and then finding a smoother pebble or a prettier shell than ordinary, whilst the great ocean of truth lay all undiscovered before me.'

ISAAC NEWTON

*Previous spread: Isaac Newton in 1684, while he was writing* Principia.

he second half of the seventeenth century saw unprecedented advances in the pure sciences, as Boyle and others got to grips with air, and Newton investigated gravity and the rainbow; important advances were also made in applied science and technology, inspired by Boyle's work with the air pump, Denis Papin applied similar ideas to steam, Thomas Savery patented an engine to raise water by the impellant force of fire, and Thomas Newcomen built a powerful steam engine that really did pump water from mines. These scientific discoveries formed the basis of the Industrial Revolution, which changed western society irrevocably, demonstrating the impact that Stuart advances had on our modern society.

# Bacon's legacy

The civil servant turned philosopher Francis Bacon inspired generations of scientists with his advocacy of hands-on science, rather than relying on traditional theory. As we have seen (page 162), he sowed the seeds of the organization that became the Royal Society: the peripatetic 'Invisible College'. From the 1640s, groups of men inclined towards the new science would meet either in Oxford or London to discuss the latest theories and developments. Their wonderfully ambitious goal was to 'make inquisitive experiments' in the hope that 'out of a sufficient number of experiments, the way of nature of workeing may be discovered'. This remains the aim of today's scientists, although the more they discover, the deeper and more fascinating the puzzle becomes. As the British physiologist J.B.S. Haldane was to write 300 years later: 'The universe is not just queerer than we imagine, but queerer than we can imagine.'

*An active proponent of experimental science, William Petty was also active in the fields of music, anatomy, medicine, surveying, and boat-designing.*

One of those early scientists, Dr William Petty, memorably summarized the appeal of experimental science: he 'never knew any man who had once tasted the sweetnes of experimentall knowledge that ever afterward lusted after the vaporous garlick and onions of phantasmaticall seeming philosophy'.

While Petty was not quite in the same league as Boyle, Wren or Newton, he is nevertheless a good example of the professional polymath: many of these early scientists were what would now be called multi-skilled. He was professor of music at Gresham and professor of anatomy at Oxford, where he 'brought back to life' a woman who had been hanged and the body given to him for dissection. In 1652 he went off as an army physician to Ireland, and then carried out a brilliant survey of that country. He designed and made simple instruments with which unskilled soldiers were able to do the work. As a result, the survey was completed for £9,000 less than the budget, and Petty kept the money. He then invented the catamaran, which proved its potential for speed in a dramatic race from Dublin Harbour to the lighthouse and back in January 1663. On the windward leg, against the tide, his boat covered two miles in 'half a quarter of an hour', which implies a speed of about 15 knots!

# Keeping time

Christiaan Huygens (pronounced 'Howgens') was a Dutch polymath who made a vital contribution to the growing science of astronomy. Born in 1629 into a grand family in The Hague, he was sent to university to study law, but spent more time on the subjects that fascinated him, mathematics and science, and by his mid-twenties was working in Paris. He and his brother built a telescope that was better than any previous instrument, and with it he observed a satellite orbiting Saturn. Galileo had spotted four moons round Jupiter forty years earlier, but this new moon, now called Titan, was entirely unexpected. (Modern astronomers are paying homage:

*The earliest Dutch pendulum clocks used a short pendulum and characteristic curved 'cheeks' at the top, to improve the regularity of the swing.*

the Cassini spacecraft at present on its way to Saturn is carrying a probe that will land on Titan – and it's called Huygens.) Huygens also recognized that Saturn's 'handles' were in fact rings around the planet – and here he went one better than Christopher Wren (see page 175). However, even the cleverest scientists can get odd ideas in their heads: he became convinced that Jupiter and Saturn were inhabited, and wrote about the ship-building and other engineering going on there.

He was on much firmer – and more productive – ground in the area of time-keeping. Astronomers have always needed to keep accurate time, to be able to measure, for example, exactly when particular stars cross the zenith (that is, reach their highest point in the sky, on the north–south line). Galileo had pointed out (page 46) that a pendulum makes a good natural time-keeper, and for some years astronomers had used pendulums to keep track of the seconds. This was inconvenient, however, because while you were watching the heavens you had to employ an assistant – usually a boy – to give the pendulum regular pushes to keep it swinging. Galileo had

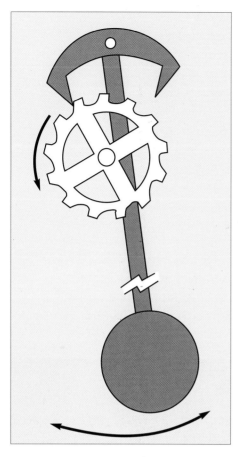

*The anchor escapement mechanism; each half swing of the pendulum tilts the 'anchor' enough to allow one tooth of the cogwheel to escape underneath.*

suggested that the pendulum could form the basis for an accurate clock, and even began trying to build one, but had never got round to completing it.

Huygens took up Galileo's idea and built the first pendulum clock in Paris in 1656. It was the most precise clock in the world, and pendulum clocks have been with us ever since. Huygens's cunning trick was to drive the pendulum using a power source – either a weight on a string or a wound-up spring instead of a boy – and to couple the power source and the pendulum to a third device that would count the number of swings, usually the hands on a clock face. This all seems obvious now, with 350 years of hindsight, but in the 1650s it was a brilliant idea and a major step forward in technology.

The critical feature of a pendulum clock – as with the verge-and-foliot clock (see page 46) – is the escapement mechanism. This mechanism couples the drive to the pendulum so that each time the pendulum swings it allows the driven cog to move forward only one tooth, and at the same time it gets a little push that keeps it swinging. The anchor escapement was another great step forward: the anchor-shaped piece of metal rocked to and fro on top of the pendulum, so that the drive wheel was stopped by its teeth, first at one end and then at the other. This contact is more precise and creates less friction than the verge system. Also the time of swing can be adjusted with great precision by varying the length of the pendulum.

Huygens tended to use short pendulums, perhaps 20 or 30 cm, but the 1-metre pendulum, taking one second for each swing, became popular for long-case 'grandfather' clocks. In 1676 one Thomas Tompion, keen to measure time precisely at Greenwich Observatory, made two clocks with 4-metre pendulums and two-second swings, using escapement mechanisms made by Richard Towneley (of whom more later). These clocks often stopped, but were the forerunners of extremely accurate observatory clocks made during the next century.

Huygens took clocks on his travels, and discovered that the force of gravity varies from place to place; near the equator the force of gravity is lower than in Paris, because the Earth bulges around the equator, so that an object on the surface there is further from the centre than one in Paris. This made his clocks run more slowly than at home. He hoped his pendulum clock would help to solve the problem of longitude (see page 203) and he took it on long sea voyages, but discovered that the rolling of the ship ruined the regularity of the pendulum's swing.

He applied his mathematical skill and worked out the relationship between the length of the string and the period of the swing; it turned out that to double the time of the swing he needed to make the string four times as long. In fact the time of swing is proportional to the square root of the length of the pendulum.

He also did work on gravity and devised a wave theory of light. He reckoned that the properties of light could be explained if light comes in waves. Isaac Newton (see below) disagreed, and was convinced that it comes in particles. Today scientists reckon they were both partly right, for light comes in units called photons, which are effectively packets of waves.

# The genius from Lincolnshire

Isaac Newton was one of the towering figures of the later Stuart age, laying the foundations of what came to be called physics. Born in Woolsthorpe Manor in 1642 – the year Galileo died – he was a weak, scrawny baby who, the midwife said, would have fitted into a quart pot (a one-litre jug); he was not expected to survive the night. He did survive, but he did not enjoy a happy childhood. His father had died before he was born, and when he was only three his mother went off to marry and live with a rich clergyman, leaving the baby with her not-very-loving parents. Eventually his stepfather

*Woolsthorpe Manor, Isaac Newton's home in Lincolnshire, with an ancient apple tree growing on the front lawn.*

'In the beginning of the year 1666... I procured me a triangular glass prisme, to try therewith the celebrated phenomena of colours.'

ISAAC NEWTON

*Opposite: Replica of Isaac Newton's reflecting telescope, made in 1671.*

died and his mother came home again, but it seems they were never close, and she sent him off to The King's School in Grantham, five miles from the family home.

Newton was probably happier at school than at home. He made model windmills and mechanical toys, which amused the other boys. He also made kites, and according to legend flew them at night with paper lanterns attached, startling the natives with these early UFOs. When his mother sent for him to go home and help on the farm he was not much use, for he was always dreaming. The story goes that, returning home one day from the market in Grantham, he got off his horse to climb Spitalgate Hill, and at the top forgot to remount, but walked all the way home, leading the horse for five miles.

His teachers persuaded his mother to send him to Trinity College Cambridge in June 1661, and four years later his genius suddenly flowered. The plague came to Cambridge – as to London – in 1665, and the University was closed down. Newton went home, and during the next eighteen months enjoyed what has come to be called his *annus mirabilis*: his miraculous year. He worked out several major outstanding problems in mathematics. He developed differential calculus – he called it the 'method of fluxions'. He sorted out the colours of the rainbow, and he began to solve the great mystery of gravity.

None of these great achievements was simple, nor free of controversy. Steps towards calculus had been taken before. Kepler, for example, when he married for the second time, laid down a cellarful of wine, and wanted to know how much he had. So he imagined each cask to be made of many thin circular slices stacked on top of one another. Then he worked out the volume of each slice and added them up. This is almost calculus. Meanwhile a German mathematician, Gottfried Wilhelm Leibniz, was also developing calculus at the same time, although quite independently. Unfortunately Newton chose not to publish his ideas for many years, and when he did so used a complicated notation, while Leibniz's was much simpler. Newton unjustly accused Leibniz of stealing his ideas, and what made things worse was that everyone on the Continent adopted the Leibniz system because it was simpler.

Newton himself described his experiments with the colours of the rainbow in a wonderful letter to the Royal Society dated 6 February 1672. This was his first publication, and he was flattered to be asked to write it. The letter is long and interesting, and reads as though he had just come from the bench after finishing the experiments. In the first paragraph he says, 'In the beginning of the year 1666... I procured me a triangular glass prisme, to try therewith the celebrated phenomena of colours.' He goes on to describe his experimental procedure in delightful detail: 'having darkened my chamber, and made a small hole in my window-shuts [i.e.

shutters] to let in a convenient quantity of the Sun's light, I placed my prism at his entrance, that it might be thereby refracted to the opposite wall.' We know which was Newton's study at Woolsthorpe; so we know exactly where he must have put the prism, and where he cast the spectrum on the opposite wall.

The current theory was that a prism produces colours because it stains the light passing through it, but Newton carefully demolished this idea. He brought up a second prism against the first, and showed that after going through two prisms 'back to back' the light came out white, whereas if each prism stained the light it should come out twice as coloured as after going through one.

Then he did what he called the *experimentum crucis* – the crucial experiment. He passed the light through one prism, and allowed the spectrum to fall on a card with a slit in it, so that only green light went through. He passed this green light through a second prism. The light was bent – it came out at a different angle – but it was still green, not stained or further coloured. Newton said this proved that the function of the prism was not to stain the white sunlight, but to split it up into its separate colours. White sunlight, he said, is actually a mixture of all the colours of the rainbow, and the prism swings the colours through different angles – red is deflected the least, blue the most.

The letter contains many other observations and little experiments, and in what is almost an aside Newton describes the reflecting telescope he had made to look at the heavens without the problem of coloured fringes. If you make a telescope with simple glass lenses you always get coloured fringes round the edges of your images, because the lenses act a bit like prisms and split up the light. This is called chromatic aberration, and it makes precise observation difficult. The problem can be avoided if you use concave mirrors instead of lenses to collect the light, because mirrors do not cause refraction, and so do not split the light into separate colours.

Newton had made his telescope in 1671, and demonstrated it to the Royal Society. It's only about 20 cm long, but still powerful enough to see the moons of Jupiter. In fact the Scottish astronomer James Gregory had designed a closely similar instrument a few years earlier, but had never made one because he could not find anyone who could grind the mirrors for him. Gregory and Newton had a fascinating exchange of letters about the respective merits of their telescopes.

However, Newton's correspondence was not all friendly and constructive. Robert Hooke, Curator of Experiments at the Royal Society, did not approve of Newton's letter about the spectrum. He claimed that the *experimentum crucis* didn't work, and that the whole theory was wrong. There ensued a bitter row that simmered on for thirty years. Newton was so angry at Hooke's caustic criticism that he said he would never publish anything again. Luckily he changed his mind about that, but he did not publish his great book on *Opticks* until 1704, a year after Hooke had died, and curiously, although he writes in *Opticks* about his work with the prism, he does not mention the *experimentum crucis*.

Isaac Newton spent a great deal of his life feuding with people who disagreed with him or who did not instantly take his side of an argument. Even his simplest statements could be barbed. In what appeared to be a conciliatory letter to Hooke he famously said, 'If I have seen further, it is by standing on the shoulders of giants.' This may just have been an acknowledgement of his debt to Aristotle, Galileo and others, but it may well have been a subtle insult. Hooke had a spinal curvature, and Newton may have been saying 'No thanks to you.' What a pity... Hooke was undoubtedly a brilliant man, and had they worked together they might have achieved even more, but it was not to be. Whenever Newton visited the Royal Society, Hooke made a point of being elsewhere, and when Newton became President of the society in 1703 he removed all traces of Hooke – even his portrait disappeared; so we do not know what Hooke looked like.

## A man of gravity

Newton's greatest achievement was in sorting out gravity. We have his own account of it from the antiquarian William Stukeley, who wrote a memoir of Newton, in which he recounts an afternoon they spent together on 17 April 1726:

*A page from the original manuscript of William Stukeley's memoir of Isaac Newton, telling the now famous story of the apple.*

> After dinner, the weather being warm, we went into the garden and drank tea, under the shade of some apple trees, only he and myself. Amidst other discourse, he told me he was just in the same situation as when, formerly, the notion of gravitation came into his mind. It was occasioned by the fall of an apple, as he sat in contemplative mood.
>
> Why should that apple always descend perpendicularly to the ground, thought he to himself? Why should it not go sideways or upwards, but constantly to the Earth's centre? Assuredly, the reason is, that the Earth draws it... there is a power, like that we call gravity, which extends itself through the universe.

Apparently Newton began to think about how far the pull of gravity extends. Clearly it reaches the top of the apple tree, to make the apple fall. Clearly it reaches the mountain tops, for snow falls on them. Could it reach as high as the Moon? If so, it must affect the Moon's orbit... and then the thought struck him. Could gravity be *causing* the Moon's orbit? He grabbed the back of an envelope and did some sums – he actually used the back of one of his mother's leasehold contracts – and said that it 'answered, pretty nearly'. In other words, in one blinding flash of inspiration he had thrown out all that nonsense about the heavenly bodies being carried around by vortices and things falling because they wanted to get back to where they belonged. A new era of science had begun.

But it wasn't quite so simple. Stevin, Galileo and others had done a good deal of work on falling bodies, and laid much of the groundwork. What is more, the story of the apple is almost certainly untrue, since there is evidence from later letters that Newton did not really develop his ideas about universal gravity until after he observed 'Halley's' comet of 1682, and did not get things sorted until he was writing his great book *Principia*, in the late 1680s. He may have made up the apple story deliberately, in order to prove to Stukeley that he had worked out the theory before Hooke – or for Voltaire, to whom he told it also – or unconsciously, in trying to work out for himself how he had arrived at the ideas about gravity.

*Title page of Isaac Newton's* Principia, *published in 1687, one of the most important science books of all time.*

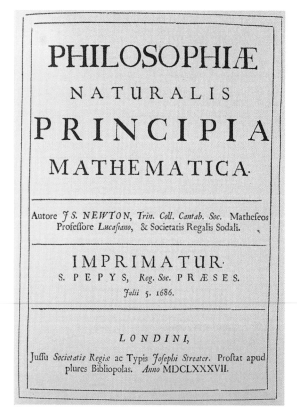

PHILOSOPHIÆ

NATURALIS

PRINCIPIA

MATHEMATICA.

Autore *JS. NEWTON*, *Trin. Coll. Cantab. Soc.* Mathefeos
Profeffore *Lucafiano*, & Societatis Regalis Sodali.

IMPRIMATUR·
S. PEPYS, *Reg. Soc.* PRÆSES.
*Julii* 5. 1686.

*LONDINI*,

Juffu *Societatis Regiæ* ac Typis *Jofephi Streater*. Proftat apud
plures Bibliopolas. *Anno* MDCLXXXVII.

Newton might never have published his work on gravity, if it had not been for Edmond Halley, one of his few close friends. Even with the help of Wren and Hooke, Halley had been unable to work out the motions of the planets, but Newton claimed he had done so. Accordingly in 1684 Halley went to visit Newton in Cambridge, and asked him for the details, and Newton said he would give Halley a paper about it. Three months later Newton started writing, and laboriously produced his masterwork *Philosophiae Naturalis Principia Mathematica* (mathematical principles of natural philosophy), generally known as *Principia*. It is a huge difficult book, written in Latin; even in English translation it is hard going. Take proposition 1 from the motion of the Moon's Nodes in Book III: 'The mean motion of the Sun from the node is defined by a geometric mean proportional between the mean motion of the Sun and that mean motion with which the Sun recedes with the greatest swiftness from the node in the quadratures.'

What a difference in style from that first letter of 1672! And yet within this great book are the laws of motion that have been used to get astronauts to the Moon. When Halley had finally prised the manuscript from Newton, he actually paid for the printing with his own money, because he believed the book was so important.

Newton went on to be Warden and then Master of the Mint; he completely recoined the country, using the milled edge on coins to make them harder to 'clip' and to forge. He became President of the Royal Society in 1703, and ruled it for 25 years with a rod of iron.

It is said that Newton had the extraordinary ability to hold a problem at the forefront of his mind and think about it continuously for long periods of time – even weeks on end. Perhaps this amazing skill stemmed from his miserable time alone as a child, and perhaps it helped him to become one of the greatest scientists of all time. At any rate, he was surely being disingenuous when, near the end of his life, he said, 'I do not know what I may appear to the world, but to myself I seem to have been only like a boy playing on the sea-shore, and diverting myself in now and then finding a smoother pebble or a prettier shell than ordinary, whilst the great ocean of truth lay all undiscovered before me.'

# Two cultures?

Newton's analysis of the spectrum produced by a prism has become a symbol of what some see as the destruction of art by science. First we had the beautiful rainbow – a symbol sent by God to promise fine weather, or in Tyndale's words, 'I do set my bow in the cloud, and it shall be for a token of a covenant between me and the earth.' Then Newton explains where the colours come from. To some this was wonderful – Alexander Pope wrote:

*Nature, and Nature's laws lay hid in night:*
*God said, 'Let Newton be!' and all was light.*

But John Keats complained that Newton had destroyed all the poetry of the rainbow by reducing it to prismatic colours:

*Philosophy will clip an angel's wings,*
*Conquer all mysteries by rule and line…*

One prominent scientist of today, Richard Dawkins, sees no such dichotomy. In the preface of his 1998 book *Unweaving the Rainbow*, in praise of the beauty and mystery of science, he wrote: 'Keats could hardly have been more wrong… Science is, or ought to be, the inspiration for great poetry…'

Erasmus Darwin, Charles's grandfather, was a doctor, an inventor, and an enthusiastic poet, whose work was much admired by Coleridge and Wordsworth. In 1791 he wrote

*Soon shall thine arm, unconquered steam, afar*
*Drag the slow barge, or drive the rapid car,*
*Or on wide-waving wings expanded bear*
*The flying chariot through the field of air.*

Darwin was inspired by the technical dreams of his friend James Watt, but the underlying scientific ideas had started building up 150 years before.

# The pressure of the atmosphere

One of the last people to work with the great Galileo before he died in 1642 was a young man called Evangelista Torricelli. Galileo had noted that even with powerful pumps he could not suck water more than about 10 metres up a pipe. He could not suck mercury up even one metre. He did not really understand this but was most interested. His pupil Torricelli picked up the idea and followed it through. He took a metre-long glass tube, closed at one end like a giant test tube, and filled it completely with mercury. Then he put his finger over the end and carefully put the end into a bowl of mercury before removing his finger. Then he tipped the tube slowly upright.

*The pressure of the atmosphere can support a column of mercury about 74cm high.*

To begin with, no mercury ran out of the tube; it remained completely full. One theory for this behaviour was that 'Nature abhors a vacuum.' If a vacuum was impossible, then mercury could not flow out of the tube,

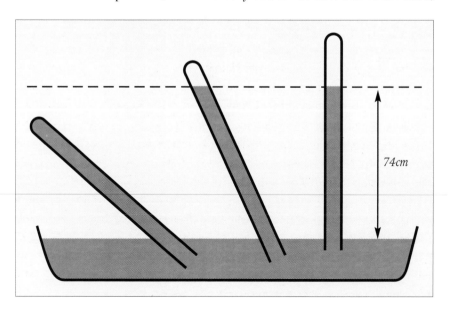

for all it could leave behind was nothing – a vacuum. As Torricelli tipped his tube further, however, there came a moment when the mercury did indeed begin to flow out of the tube, leaving a gap above its surface. There did not seem to be anything in this gap, since when he tilted the tube down again the mercury went right to the end, which it would not if the gap was full of air. So Torricelli seemed to have made the impossible vacuum, and it is sometimes called a 'Torricellian vacuum'.

By careful observation he noted that what really mattered was not the angle of the tube, but the vertical height of the column of mercury; the mercury level in the tube would never rise more than about 74 cm above the mercury surface in the bowl. And then he realized what was going on. We live at the bottom of a sea of air, and the air has weight; so it presses on us with what is called atmospheric pressure. Atmospheric pressure is about 1 kg on each square centimetre (or 15 pounds on each square inch), and this is just enough to hold up a column of mercury 74 cm high, or a column of water 10 metres high.

*Evangelista Torricelli, a pupil of Galileo and the inventor of the barometer.*

In Oxford, Boyle and Hooke discovered that the height of this mercury column seemed to depend on the weather; they got a slightly higher column in fine clear weather than on wet days, and it was lowest during storms. Because the height of the column seemed to change before the weather did, they thought this might be a way of predicting the weather; in fact Torricelli had invented the barometer. Early barometers were simply Torricellian tubes full of mercury, calibrated so that you could read off the atmospheric pressure.

## The Magdeburg hemispheres

Torricelli's work caused masses of argument; philosophers still said a vacuum was impossible, and he must be making a mistake. In Magdeburg, however, the German scientist Otto von Guericke was excited by Torricelli's ideas, and began to study the properties of air; in 1650 he invented the first air pump. In 1654 he laid on a superb demonstration, famous to this day. He had made two bronze hemispheres, which fitted together round their circumference to make a hollow sphere of about 50 cm diameter. When just put together, the hemispheres fell apart; however, after he had pumped out the air using his air pump, two teams of horses were unable to pull them apart – to the astonishment of the assembled dignitaries.

Were the horses feeble? Was von Guericke cheating? Probably not. If he had been able to pump out all the air, and make a perfect vacuum inside,

*The Magdeburg hemispheres. Otto von Guericke demonstrated that once the air was pumped out, the hemispheres could not be pulled apart by two teams of horses.*

the pressure of the atmosphere on the outside of each hemisphere would have held them together with a force of about two tonnes. In other words, even with the partial vacuum that was the best he could make with his pump, the atmospheric forces were more powerful than the horses.

News of this demonstration buzzed through the scientific community, reaching the ears of a certain pioneering Irishman: Robert Boyle.

# Boyle's Law – or
# Mr Towneley's hypothesis

Born in Lismore Castle on the south coast of Ireland in 1627, Robert Boyle was the seventh son of the First Earl of Cork. He was brilliant – hailed as 'the father of modern chemistry' – though neurotic and a hypochondriac. While he was studying in Venice he heard about the death of Galileo, and was moved to read some of Galileo's books. He jumped at Bacon's ideas of hands-on science, and quickly began doing experiments.

When he heard about von Guericke's air pump, Boyle realized that it would be a powerful weapon for investigating the properties of air. So he hired a young man called Robert Hooke to make one for him. Hooke (see page 165) was an enormously ingenious man, and remained Boyle's assistant for some time; indeed Boyle was not particularly good with his hands, and had terrible eyesight, and some people say that Hooke not only did most of the experiments but also wrote them up for Boyle.

With his pump, Boyle was able to pump the air out of a large glass jar and investigate the properties of the resulting vacuum. His pump was simple, and therefore his vacuum must have been far from perfect, but a partial vacuum was quite good enough to show several things:

> *A candle was placed burning in the jar; the flame went out as the air was pumped away.*

> *A watch was set ticking in the jar; when the air had been pumped out the ticking of the watch could no longer be heard.*

> *A small bird was put in the jar. When the air was pumped out the bird died.*

From these experiments Boyle deduced various properties of air: that it is necessary for combustion, for the transmission of sound, and for life – animals need air to breathe. He showed that if he simply pumped out the air and then opened the valve, the air rushed back in with a loud hiss. He did not have to push it in – it went of its own accord. These simple experiments incidentally provided powerful support for Bacon's ideas: Boyle was learning by investigation a range of things that would have been difficult if not impossible to deduce by argument.

He put in his jar a lamb's bladder, sealed, with just a little air inside. Then he pumped the air from the jar, and the bladder expanded. He called this 'the spring of the air' and imagined air as being made of a large number of little pieces like little tight and springy coils of wool, ready to expand if given the chance.

He was most interested to hear from his friend Richard Towneley of some related work. Towneley was the first person in the country to measure and record rainfall in a systematic way. He put a collecting cylinder on the roof of his home, Towneley Hall near Burnley in Lancashire, and piped the rainwater down into the house for measurement. For twenty-five years he recorded the rainfall three times a day, and discovered that the weather was much wetter in Lancashire than in Paris, because, he thought, of the high ground in east Lancashire and Yorkshire; the prevailing south-west winds brought in rain clouds which 'are oftener stopt and broken and fall upon us'.

*The first systematic measurement of rainfall in England was carried out by Richard Towneley at his home, Towneley Hall, near Burnley in Lancashire.*

On 27 April 1661 Towneley collected a sample of 'valley ayr' at Towneley Hall, and carried it with the help of an assistant to the top of Pendle Hill, a stiff climb of about 300 metres. Up at the top he found that the volume of the air had increased. There he collected some 'mountain ayr' which he took down with him, and found its volume decreased. He believed this was because the pressure was lower at the top of the hill, and he speculated in his letter to Boyle about whether there might be an inverse relationship between volume and pressure: the lower the pressure the greater the volume, and vice versa.

Boyle was intrigued, checked it with his air pump, and decided that Towneley was right. He wrote it up the following year, and the relationship between pressure and volume came to be called Boyle's Law. However, Robert Boyle himself called it 'Mr Towneley's hypothesis'.

Boyle became well-known for his interest in air and gases, and as a result received some years later a wonderful letter from John Clayton, rector of Crofton church near Wakefield, who discovered and experimented with coal gas: 'Having seen a ditch within two miles from Wigan in Lancashire, wherein the Water would seemingly burn like Brandy, the Flame of which was so fierce, that several Strangers have boiled Eggs over it; the people thereabouts indeed affirm, that about 30 years ago it would have boiled a Piece of Beef...'

He dammed the ditch, baled out the water, and dug down: 'about the Depth of half a Yard, we found a shelly Coal, and the Candle being then put down into the Hole, the Air catched Fire, and continued burning'. He took some coal home with him, distilled it 'in a Retort in an open Fire', and made a gas that he called 'Spirit of Coal', which he kept for some time in bladders, and amused his friends by lighting with a candle and burning in jets. A hundred and fifty years later, this coal gas would light the homes and streets of Victorian England.

The properties of air had also been studied in an entirely different context: with the aim of enabling people to breathe under water, in a submarine or diving bell.

# Making oxygen?

An ingenious Dutchman called Cornelius Drebbel was probably the first to build an effective submarine. In fact he was fascinated by all the elements – earth, fire and water, as well as air – and wrote a book about them. Born in Alkmaar in 1572, he was another polymath, being painter, engraver and alchemist too. Apart from the submarine, he invented all sorts of things from a pump to a clockwork mechanism that would allegedly run for a hundred years.

Drebbel came to England around 1605 at the invitation of King James I, and was given a pension and lodging at Eltham Palace near Greenwich. With him he brought the idea of a compound microscope, which was to be so useful in the hands of Robert Hooke (see page 165), and also of a telescope, although Digges probably made one some time earlier (see page 87). He devised a still for obtaining fresh water from salt water, wind-powered musical instruments, an instrument for 'showing pictures of people not present', and a 'perpetual motion machine' – probably driven by variation in atmospheric pressure. This told the time, date and season, and made Drebbel so famous that in 1610 he was summoned to Prague to demonstrate it to Rudolph II, the Holy Roman Emperor. Unfortunately Rudolph was deposed by his brother, and Drebbel was imprisoned until the Prince of Wales intervened, when he was allowed to return to England.

Drebbel built his submarine around 1620. It seems to have been like two rowing boats, one turned upside-down and placed on top of the other. The whole vessel was encased in greased leather to keep it watertight. Mark I was small, with just four oars – rowing under water must have been tricky, but possible. The crew had bladders under their seats, connected by tubes to the outside. When they wanted to submerge they opened the tubes (by untying them) to let water in. This raised the density of the vessel and it sank. When they wanted to surface again they squashed the bladders flat to push out the water, and then tied off the tubes to keep the water out.

Apparently the submarine could cruise along four metres below the surface and, it is said, travelled under water in the River Thames from Westminster to Greenwich and back in three hours. Since this is a round

*Cornelius Drebbel's submarine being demonstrated to James I about 1620; note the old St Paul's on the opposite bank.*

trip of some six miles, the story is hard to believe – especially as the rowers would not have been able to see where they were going – but the submarine really does seem to have worked and to have been demonstrated to the king, who wisely declined to have a trip under water. Drebbel went on to build bigger submarines: Mark 3 had twelve oars and could allegedly carry sixteen people (which must have been rather a squash). This vessel was strengthened with iron bands, and even had windows.

There is a complete mystery about Drebbel's air supply. According to one account, there were pipes to the surface and bellows to pull air down, but Robert Boyle spoke many years later to one of the people who had been on board, and Boyle wrote that Drebbel had a 'chemical liquor' that would replace the 'quintessence of air' that could 'cherish the vital flame of the heart'. In other words, Drebbel seems to have had a supply of oxygen, or something that produced oxygen, about 150 years before it was officially discovered by Joseph Priestley, Antoine Lavoisier and Carl Wilhelm Scheele.

## *Halley's diving bell*

In about 1690 Edmond Halley, not content with being 'the Southern Tycho', the man who worked out the path of the eponymous comet, and the man who persuaded Isaac Newton to write the most important textbook of all time (see page 218), invented an improved diving bell, as he described in a paper published in the *Philosophical Transactions* of the Royal Society. Approaching the challenge like a true scientist, he first explains the main problem:

> *Since by Experiment it is found that a Gallon of Air, included in a*
> *Bladder, and by a Pipe reciprocally inspired and expired by the Lungs*
> *of a Man, will become unfit for any further Respiration, in little more*
> *than a minute of Time, and though its Elasticity be but little altered,*
> *yet in passing the Lungs, it loses its vivifying Spirit, and… in an*
> *instant extinguishes the brightest Flame… a naked Diver… may not be*
> *above a couple of Minutes enclosed in Water… without Suffocating.*

He constructed a diving bell like an upturned wooden bucket, three feet in diameter at the top and five at the bottom, weighted down with lead, and with a glass window in the roof to let in light. He explains in his paper that as it is lowered, the water compresses the air inside: 'This included Air, as it descends lower, does contract itself according to the weight of the Water that compresses it; so at thirty three Foot deep or thereabouts, the Bell will be half full of water, the Pressure of it being then equal to that of the whole Atmosphere.'

He seems to have gone down himself and remained 'nine or ten fathoms' (18 metres/60 feet) under water for 'above an Hour and a half at a time, without any sort of ill consequence… Besides the whole Cavity of the Bell was kept entirely free of Water, so that I sat on a Bench, which was diametrically placed near the Bottom, wholly dressed with all my Cloaths on.' He describes vividly the experience of going down: 'The only inconvenience that attends it, is found in the Ears… on the first Descent of the Bell, a Pressure begins to be felt on each Ear, which by Degrees grows painful, like as if a Quill were forcibly thrust into the Hole of the Ear.' However, he says, coming up again is much less painful, and there seems to be no permanent damage to the 'Organs of Hearing'.

His great improvement over earlier diving bells lay in his mechanism for getting more air down to the bell, both to push out the water that comes in at depth, and to refresh the air for the diver to breathe. Many neat inventions seem blindingly obvious with the benefit of hindsight; in a charmingly self-effacing way Halley says of his method of getting air down: 'This I effected by a Contrivance so easy, that it may be wondred it should not have been thought of sooner.'

He covered two barrels with lead so that they would sink even when empty, left a bung-hole open at the bottom of each, and in the top fixed an oiled leather hose long enough to hang down to the bottom of the barrel. Each of these was lowered in turn from the air above the water

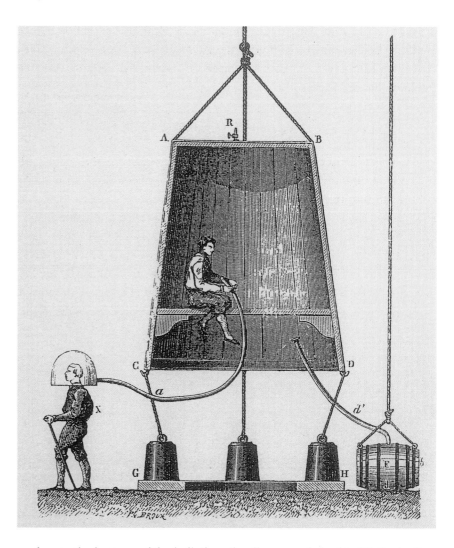

*Edmond Halley's diving bell being supplied with fresh air via a hose from the barrel on the right.*

surface to the bottom of the bell; then the diver merely had to lift the leather hose inside the bell, and all the air would flow up from the barrel into the bell. He used these barrels to top up the air during his descent, and so keep the bell empty of water, and then when he was settled on the bottom to refresh the air every few minutes, letting out an equal amount of stale air through a stopcock at the top of the bell.

He proposed that such diving bells might be used for pearl fishing, diving for sponges and coral, working on bridges and other underwater constructions, and scrubbing of ships' bottoms; however, the generally accepted theory is that he built his bell primarily to try and recover sunken treasure. He was no doubt encouraged by the news that one William Phipps had used a diving bell in the West Indies to recover £200,000 in treasure from sunken Spanish galleons; Charles II gave Phipps £2,000 and a knighthood.

The French scientist Denis Papin had proposed in 1689 that diving bells could be provided with air from the surface by force pumps. Halley must have known of this suggestion, but avoided trying it, either because he did not believe the force pumps then available were up to providing all that air under pressure, or because he did not want to be seen to be pinching Papin's ideas.

# The impellant force of fire

The work of Torricelli and Boyle had showed that air could be pumped, manipulated and investigated. Next in line was steam, and Denis Papin was one of the first people to put steam to serious use. Hero of Alexandria had made a toy steam engine hundreds of years earlier, and Giambattista della Porta (see page 78) had written in 1606 about the idea using 'the impellant force of fire', but Papin wanted to make steam do some real work.

Papin was born at Blois in the Loire Valley and studied to be a doctor, but went to Paris to work with the great Christiaan Huygens, who was by

*Denis Papin invented the pressure cooker in 1680.*

then Director of the Académie des Sciences, and was following up the work of von Guericke and Boyle on the air pump. Papin made significant improvements to Huygens's air pump, and when he travelled to London in 1675 he had a letter of introduction to Robert Boyle, which gave him immediate access to the Royal Society. Papin became temporary Curator of Experiments there in 1679, although like many others he found it difficult to get on with Robert Hooke.

In 1680 he invented 'A New Digester or Engine for Softening Bones' – in other words a pressure cooker. He understood that if you boiled water under pressure the boiling point would increase, which would reduce cooking times. He carried out a long sequence of experiments to investigate the effects, and because no good thermometer had been invented he estimated the temperature of his pressurized saucepan by putting a drop of water in a dent in the lid and seeing how many times a pendulum swung before it evaporated.

In experiment 11 on 13 July he took an 'old male and tame Rabbet, which is ordinarily but a pitiful sort of meat', and burned six ounces of coal in his fireplace, by which time the water drop disappeared in four seconds. He reckoned that the pressure inside the digester was about six times stronger than normal air pressure – which is a bit terrifying, since a modern pressure cooker would raise the pressure by less than two times.

However, the result of experiment 11 was a great success: the Rabbet was well cooked, the bones softened, and it tasted as good as young Rabbet. He was particularly pleased that the juice made good 'gelly'. He was most enthusiastic about gelly, described how to make it from softened bones, and claimed that gellys are good for several diseases. On 12 April 1682 Papin used his digester to cook an entire 'philosophical supper' for the President and Fellows of the Royal Society. Sadly, the menu does not survive, although the pressure cooker lives on.

Papin was keen to make an engine, and after considering one powered by gunpowder he settled in 1690 for a steam engine, or rather an atmospheric engine. The cylinder was filled with steam to push up the piston. The steam was then condensed to make a vacuum, and the pressure of the atmosphere pushed the piston down again. It was sometimes called an atmospheric engine because the main driving force was supplied by the pressure of the atmosphere; Papin was building on his experience with the air pumps. He described clearly how it would work:

> Since it is a property of water that a small quantity of it turned into
> vapour by heat has an elastic force like that of air, but upon cold
> supervening is again resolved into water, so that no trace of the said
> elastic force remains, I concluded that machines could be constructed

*wherein water, by the help of no very intense heat, and at little cost, could produce that perfect vacuum which could by no means be obtained through gunpowder.*

Dismissing this engine as being too inefficient, Papin began to develop a high-pressure steam engine, in which the work would be done not by the atmosphere but by the pressure of steam at high temperature. Alas, this pioneering work failed to make Papin rich and famous, because he did not believe in patenting his ideas – rather, he wanted to share them with the world to allow maximum benefit to be derived. Others were less altruistic, and the man who jumped on to what he perceived to be a bandwagon was Thomas Savery, of Shilston Barton near Modbury in Devon.

Savery was a younger son who did not inherit land, and needed to earn his living. He seems to have been a military engineer, and he described a machine for making glass marbles and one for propelling a ship – a sort of pedallo. He had access to the society of London, and may well have heard of Papin's work, although it is possible that he came up with the idea on his own. What seems likely is that living close to Cornwall he had heard of the terrible problems of the flooding in Cornish mines. They had been worked since Roman times, and had become deeper and deeper, and steadily filled with water. At first men pumped the water out by hand; then they used horses; but in the deeper mines horsepower was not enough. They needed a more powerful engine.

Thomas Savery saw the potential; a steam engine might be able to pump water from the mines, and there was money to be made. Fortunately he knew the king – William III – and in 1698 was able to take out a patent for an 'Engine for Raiseing Water by the Impellant Force of Fire' (presumably he had picked up the phrase from della Porta). His patent has no diagram of his engine, nor even a description, but he clearly meant it to succeed, for in 1702 he produced a book called *The Miner's Friend*, which was a combined sales brochure and instruction manual. It starts with a grovelling letter to the king, in very large print – perhaps he thought William had poor eyesight – followed by a letter to the Gentlemen of England describing the benefits to be gained by using his engine.

The instruction section is clear and detailed. A boiler provided steam which filled the working cylinder. The supply of steam was then cut off, and a valve opened to connect the steam to the water in the mine. This cold water condensed the steam, creating a vacuum, whereupon the pressure of the atmosphere pushed water up the pipe into the cylinder. When it was full of water, the lower valve was closed, the upper valve opened to allow the water up to a higher level – say 10 metres above the engine – and the supply of steam reconnected. Then the high-pressure steam pushed the water up to the next level – ideally out of the mine.

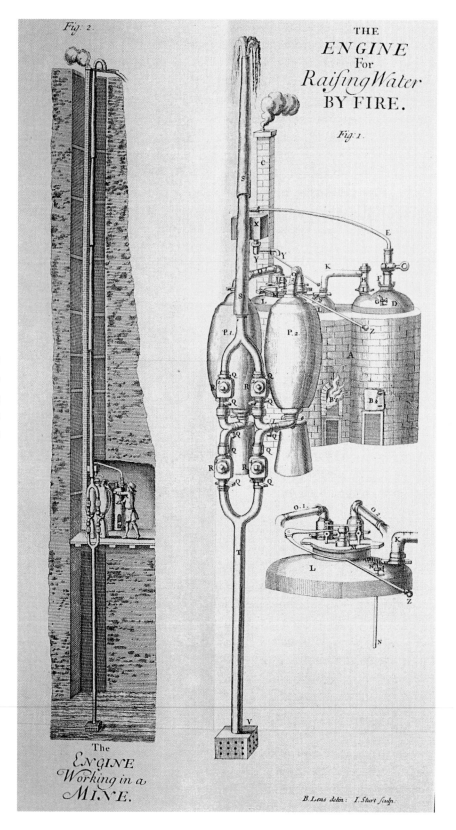

*A diagram from Thomas Savery's book* The Miner's Friend, *showing how his steam engine should be installed to pump water from a mine.*

Savery suggested placing his engine half-way down the mine shaft, so that it could pull water up from the bottom (less than 10 metres) and then push it the rest of the way up the shaft – perhaps a further 10 metres. In mines more than 20 metres deep there would have to be engines stacked about every 20 metres up the shaft – and some mines were 500 metres deep… He was at pains to say how powerful his engine was, and he actually invented the term 'horsepower' – deliberately making it rather more than most horses could manage so that his customers would not be disappointed.

Savery's steam engines never worked well. A few were probably used in mines; one was installed to control the water supply at Hampton Court; another at Campden House in Kensington was still running eighteen years later. But there was a real problem. The water was pushed out by positive steam pressure, and that needed high-pressure steam. The solder and the joints could not cope, and so the machines kept going wrong. Savery did not get rich until the very end of the Stuart period, on the proceeds of another man's ideas.

# The common engine

Thomas Newcomen was an ironmonger in Dartmouth, a serious-minded and religious artisan who made and repaired tools for workers in many local industries, including the miners of Cornwall. He too knew all about the problems of flooding in the mines, and must have been looking for solutions. Some time in the early 1700s he designed an atmospheric engine similar to the one described by Papin ten years earlier, although it seems unlikely that a Devon ironmonger should have had access to ideas floating about in London.

Newcomen's engine may have been the greatest ever single step forward in technology, because it represented mobile power. Before then you needed human or animal muscle-power, or you needed access to a waterwheel in a river or a windmill. With a Newcomen engine, or a 'common engine' as it came to be called, you could have unlimited power anywhere and at any time. Newcomen's engines had a piston in a single cylinder, supplied with steam by a boiler underneath. They were designed specifically for pumping water out of mines, and they were installed in purpose-built engine-houses, tall buildings with high chimneys to improve the draught for the boiler furnace.

Above the cylinder was one end of a massive beam, which was supported and pivoted on one wall of the engine house. From the other end, stuck outside the engine house, heavy iron pumping rods hung straight down into the mine. When the steam valve was opened, allowing steam into the cylinder, the weight of the pumping rods pulled down the outside end of the beam; the inner end rose, lifting the piston, and the cylinder filled

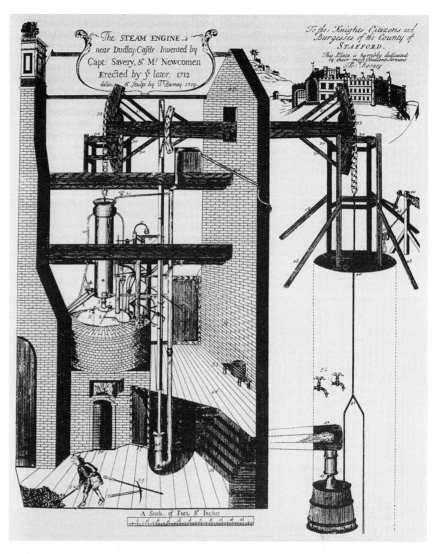

The Newcomen engine, or Common engine, was arguably the greatest ever single step forward in technology.

with steam. Then the steam valve was shut and cold water squirted into the cylinder to condense the steam, and create a partial vacuum inside. The pressure of the atmosphere then forced the piston down into the cylinder, lifting the pump rods and so pumping the water upwards. The pump rods were attached either to a force pump to push the water up, or simply to a train of buckets to lift it, one level at a time.

Like Papin's first model, this was an atmospheric engine – getting its power stroke from the force of atmospheric pressure. The two important innovations were that the engine used a piston in a cylinder to create the power stroke, and used a force pump or mechanical lifting, rather than sucking up water, which was what Savery's engine did. The first cylinders were fairly small – only 50 or 60 cm in diameter – and were made from brass, but later ones were made from cast iron, which was cheaper, and grew to vast sizes – 2 or even 3 metres in diameter and 3 or 4 metres high.

The first recorded Newcomen engine was installed at a coal mine at Dudley in the West Midlands in 1712 and caused quite a stir. People came from far and wide to see this great wheezing monster that pulled hundreds of gallons of water from the mine every hour, day in, day out. The potential was obvious, and soon orders were pouring in. Few could afford to buy one of these engines – they cost about £1,000 each – so Newcomen arranged for the building and then took a royalty of around £1 a day, or 35 per cent; so he could recover his money in three years. Unfortunately for him he did not get rich, because his engines were covered by Savery's all-embracing patent for 'an engine for raiseing water by the impellant force of fire', and Savery had managed to get his patent extended until 1725. Newcomen and Savery probably went into partnership.

For all its power, the Newcomen engine was inefficient; it used huge amounts of coal to keep raising steam. This did not matter too much in coal mines, where there was plenty of coal about, and coal of the poorest quality could be used for the boiler. However, the inefficiency did matter in the Cornish mines. There was no coal in Cornwall, and importing it from South Wales to run the steam engines was too expensive. So ironically, even though both Savery and Newcomen had developed engines for them, the Cornish miners had to wait a further sixty years, until James Watt made his superb improvements, before they could afford steam engines to drain their mines.

Nevertheless, in spite of its inefficiency, the common engine was a phenomenal success. More than a hundred were built before Thomas Newcomen died in 1729, and more than a thousand were built eventually. They were exported all over the world; the idea was taken to America by Joseph Hornblower, and some Newcomen engines ran for more than a hundred years.

This was the beginning of the Industrial Revolution, and a turning point for the human race. The dreams of Francis Bacon had come true, for scientific experiments had brought about understanding and changes to the benefit of mankind. Isaac Newton had sorted out gravity and the rainbow. More specifically, Bacon had called for empirical experiment and the formation of a scientific academy. Robert Boyle and others had started such an academy, which turned into the Royal Society, and Boyle had done experiments with his air pump on the 'spring of the air'. Papin, Savery and finally Newcomen used those ideas about the spring of the air to build steam engines which began the process of transforming the western world through the use of technology. The best part of a hundred years would go by before steam engines could drive cotton mills, weaving machines, ploughs and motor cars, and yet all these things were enabled by the first practical common engines of Thomas Newcomen. This was the ultimate legacy of the Stuarts: the appliance of science.

# Timeline

1476   William Caxton brings to England printing with moveable type

1485   Henry VII comes to the throne

1492   Christopher Columbus sails west and discovers Caribbean islands

1496   First dry dock built at Portsmouth

1496   First English blast furnace built in the Weald

1497   **John Cabot** sails from Bristol and discovers mainland North America

1497/8   Vasco da Gama sails round Africa to India

1498   Toothbrush invented in China

1504   First coins with likeness of monarch – Henry Tudor

1507   Martin Waldseemüller's world map shows America for first time

1509   **Henry VIII** becomes king, marries Catherine of Aragon

1510   Gun-ports invented, warship *Mary Rose* built

1510   First copyright in Britain granted to Thomas Godfry

1513   Henry VIII invades France

1517   Martin Luther begins protestant movement

1519-21   Magellan (or at least his ship) circumnavigates world

1525   William Tyndale publishes **New Testament** in English

1530   Tyndale publishes Genesis in English; his New Testament is banned in England

1532   Henry VIII declares himself Head of the Church of England

1533 Henry VIII marries Anne Boleyn and is excommunicated
1535 Sir Thomas More beheaded
1535/6 Henry VIII dissolves monasteries
1536 William Tyndale executed for heresy
1538 Henry VIII puts an English Bible in every church
1537 Jane Seymour dies
1540 Henry marries Ann of Cleves
1543 Nicolaus Copernicus publishes heliocentric theory
1543 Andreas **Vesalius** publishes *De Humani Corporis Fabrica*
1543 Ralph Hogge casts first complete iron **cannon**
1544 Southsea Castle built
1545 French try to invade Isle of Wight; Mary Rose sinks
1546 First civil divorce in Britain
1547 Henry VIII dies; Edward VI comes to the throne
1550 Girolamo Cardano describes camera obscura with a lens
1553 Edward VI dies; Mary becomes Queen
1556 *De re metallica* published
1557 Robert Recorde invents the equals sign
1558 Mary dies; **Elizabeth** becomes Queen
1559 Anthosis van der Wyngaerde's street map of London
1568 Gerhardus Mercator produces projection of map of world
1571 Leonard Digges describes the theodolite, and perhaps a telescope
1571 Royal Exchange so named by Queen Elizabeth
1572 Supernova observed by **Tycho Brahe** in Denmark, also by John Dee and Thomas Digges in England
1576 First theatres built in London
1577-80 **Drake** sails round world, stealing Spanish gold
1582 First water piped to private houses in London
1583 Galileo describes regularity of pendulum swing
1584 First pencils made with graphite from Borrowdale
1585 Sir Walter Ralegh sets up first British colony in North America
1586 Simon Stevin drops things from leaning tower in Delft
1586 Sir Thomas Harriot brings first potatoes to Britain (?)
1587 Mary Queen of Scots executed after Sir Francis Walsingham uncovers Catholic plot

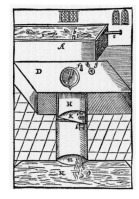

1588 **Spanish Armada** routed by English navy

1588 Elizabeth I issues standard weights and measures

1589 Rev. William Lee invents knitting machine

1589 First paper mill in England

1589 Galileo drops things from leaning tower of Pisa

1591 **William Shakespeare** begins acting and writing plays

1592 First thermometers (Galileo and others)

1596 John Harington describes his Ajax **water-closet**

1600 Giordano Bruno burned at the stake for heresy

1600 William Gilbert publishes *De Magnete*

1603 Fabrizio observes valves in veins

1603 Elizabeth I dies – end of Tudor dynasty

1603 James VI of Scotland becomes also James I of England

1605 Gunpowder plot foiled

1606 First union flag combines flags of St George and St Andrew

1607 Sir Thomas Chaloner granted patent to make alum

1608 Hans Lippershey invents telescope

1608 **Inigo Jones** turns to architecture

1608 Table forks introduced from Italy by Thomas Coryate

1609 Johannes Kepler publishes laws of planetary motion

1609 Lippershey & Jansen make compound **microscope**

1609 Hugh Myddelton begins digging New River

1609-10 Galileo looks at sky with telescope

1610 Flintlock gun invented

1614 John Napier publishes book of logarithms

1617 Napier's bones – the first pocket calculator

1617 Rathburne & Burgess granted patent No 1 for street maps

1620 **Francis Bacon** publishes *Novum Organum*

1620 Cornelis Drebbel demonstrates the first **submarine**

1624 Henry Briggs calculates logarithms to base 10

1625 James I dies; Charles I becomes King

1626 Francis Bacon dies inventing frozen chicken

1628 William Harvey publishes his book on circulation of blood

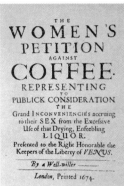

1632 Galileo publishes *Dialogue concerning the two world systems*

1633 Bananas introduced to Britain in a shop window in London

1637 Descartes' analytical geometry

1637 Pierre de Fermat dreams up his 'last theorem'

1637 First recorded use of waterproof umbrellas in France

1639 Jeremiah Horrocks observes transit of Venus

c. 1640 William Gascoigne's micrometer for measuring moon

1642 Galileo dies; Isaac Newton in born

1642 Press freed from censorship

1642 Civil war in England

1642 Blaise Pascal makes mechanical calculator

1643 Captain Baily invents the hackney cab

1644 Evangelista Torricelli invents barometer

1646 Pascal shows air pressure is lower up mountains

1647 **Cromwell** reintroduces censorship of the press

1649 Charles I executed

1649 Sir Ralph Verney asked to bring toothbrushes from Paris

1650 Otto von Guericke invents the air pump

1650 The first **coffee-house** opened in England

1650 Marriage bureau opened in London by Henry Robinson

1651 Harvey's embryology - *On the generation of animals*

1651 Tea first sold to the public by Thomas Garway

1654 Pascal invents probability theory to help gambler

c. 1655 John Ray classifies plants, inventing species

1656 Christiaan Huygens invents pendulum clock

1657 First **coach service** introduced, from London to Chester

1658 Robert Hooke invents watch spring

1660 Boyle's Law derived from Towneley's Hypothesis

1660 Restoration of **Charles II**

1660 Royal Society formed

1661 Robert Boyle defines an element in *The sceptical chemist*

1661 John Evelyn writes about air pollution

1661 Malpighi describes capillaries

1662 Charles II marries Catherine of Braganza

1662 William Petty invents the catamaran in Dublin

1663 James Gregory invents reflecting telescope

1663 Margaret Hughes is one of first women to appear on stage

1665 Robert Hooke publishes *Micrographia*

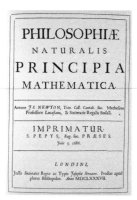

1665 Great Plague

1666 Great Fire of London

1666 **Isaac Newton**'s *annus mirabilis*

1666 Samuel Morland invents an adding machine

1668 Newton invents reflecting **telescope**

1669 **Christopher Wren** appointed Surveyor General

1669 Hennig Brandt isolates phosphorus

1670 Samuel Morland describes his *Tuba stentorophonica* (megaphone)

1671 Jean Picard measures size of the earth

1671 Anthony van Leeuwenhoek begins microscopic work

1672 Newton's first publication, on the spectrum and telescope

1673 Giovanni Domenico Cassini estimates distance from Earth to Sun within 7 per cent

1675 Olaus Roemer calculates speed of light

1675 John Flamsteed appointed (first) Astronomer Royal

1675 Christopher Wren builds **Greenwich Observatory**

1675 John Ogilby publishes first road atlas

1677 Richard Towneley begins to record rainfall

1679 Christiaan Huygens describes wave theory of light

1680 Denis Papin invents the pressure cooker

1680 Edward Lloyd opens coffee-house, and insurance business

1682 Edmond Halley observes a comet

1683 William Dockwra establishes penny post

1685 Charles II dies; James II becomes King

1687 John Clayton, rector of Crofton, discovers coal gas

1687 Newton's **Principia** published at Halley's expense

1688 William and Mary take over throne in 'glorious revolution'

1690 Denis Papin's steam engine

1690 Edmond Halley describes an improved **diving bell**

1693 Publication of *Ladies' Mercury*, with agony column

1694 Bank of England opened

1695 Sir Samuel Morland erects first public drinking fountain

1698 Thomas Savery's steam engine patent

1698 Henry Winstanley's first Eddystone Lighthouse

1700 Chatsworth has bathroom with hot and cold running water

1701 Jethro Tull invents the **seed drill**

1702 William III dies; his wife's sister Anne becomes Queen

1702 *Daily Courant* – first daily newspaper

1703 **Eddystone Lighthouse** swept away by great storm

1704 Newton publishes *Opticks*

PHILOSOPHIÆ

NATURALIS

PRINCIPIA

MATHEMATICA

Autore JS. NEWTON, Trin. Coll. Cantab. Soc. Mathefeos Profeffore Lucafiano, & Societatis Regalis Sodali.

IMPRIMATUR·
S. PEPYS, Reg. Soc. PRÆSES.
Julii 5. 1686.

LONDINI,

Juffu Societatis Regiæ ac Typis Jofephi Streater. Proftat apud plures Bibliopolas. Anno MDCLXXXVII.

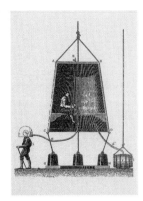

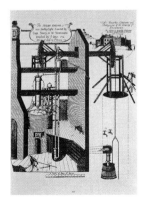

1704 Benjamin Huntsman invents crucible steel

1705 **Halley** predicts the return of his comet

1708 St Paul's Cathedral completed

1707 Act of Union between England and Scotland

1707 Sir Clowdisley Shovell crashes into Scilly Islands

1709 Abraham Darby discovers how to make iron using coke

1709 First piano built by Bartolommeo Cristofori in Florence

1711 John Shore produces tuning fork in London

1711 Royal Ascot instituted by Queen Anne

1712 Thomas Newcomen builds his first **steam engine**

1714 Anne, the last Stuart, dies; her second cousin becomes George I

# Bibliography

Agricola, Georgius, *De Re Metallica 1556*
  (trans. Hoover & Hoover, Dover Publications Inc., 1950)
Bolles, Edmund Blair, *Galileo's Commandment*
  (Little, Brown and Company, 1997)
Cajori, Florian (ed), *Sir Isaac Newton – Principia*
  (University of California Press, 1934)
Cardwell, Donald, *The Fontana History of Technology*
  (Fontana Press, 1994)
Carey, John, *The Faber Book of Science* (Faber and Faber, 1995)
Chambers, James, *Christopher Wren* (Sutton Publishing, 1998)
Crystal, David, *The Cambridge Encyclopedia* (Guild Publishing, 1991)
Daniell, David, *William Tyndale* (Yale University Press, 2001)
Derry, T. K. and Williams, Trevor I., *A Short History of Technology*
  (Oxford University Press, 1960; Dover Publications Inc., 1960)
Devey, Joseph (ed), *The Physical and Metaphysical Works of Lord Bacon*
  (George Bell & Sons, 1904)
*Dictionary of National Biography* (Oxford University Press, 1982)
Emsley, John, *Nature's Building Blocks* (Oxford University Press, 2001)
*Encyclopaedia Britannica* 9th ed., (Adam & Charles Black, c. 1875)
*Dictionary of World History* (Chambers, 1993)
Erickson, Carolly, *Great Harry* (Robson Books, 1998)
Fraser, Antonia, *Cromwell*, (Phoenix Press, 1973)
Fraser, Antonia, *King Charles II* (Arrow, 1992)
Fraser, Antonia, *The Gunpowder Plot* (Arrow, 1996)
Fraser, Antonia, *The Lives of the Kings & Queens of England,*
  (Cassell & Co, 1975)
Hardy, Robert, *Longbow* (Patrick Stephens Limited, 1976)
Hart-Davis, Adam, and Troscianko, Emily, *Henry Winstanley and the
  Eddystone Lighthouse* (Sutton Publishing, 2002)
Hart-Davis, Adam, *Chain Reactions,* (National Portrait Gallery, 2000)
Hart-Davis, Adam, *100 Local Heroes* (Sutton Publishing, 1999)
Hart-Davis, Adam, *Thunder, Flush and Thomas Crapper*
  (Michael O'Mara Books Ltd.,1997)
Hooke, Robert, *Micrographia (1665)*
Inwood, Stephen, *A History of London* (Macmillan, 1998)
Kollerstrom, Nicholas, 'The Path of Halley's Comet, and Newton's Late
  Appreciation of the Law of Gravity', *Annals of Science 56* (1999)
  pp.331-356

Marshall, Gary, *The Ravenscar Alum Works*,
  (Scarborough Archaeological and Historical Society, 1990)

Martin, Steve and Sanger, Colin, *Matthew* (Godrevy Publications, 2000)

McNeil, Ian, *An Encyclopaedia of the History of Technology*
  (Routledge, 1990)

McNeil, Ian, *Biographical Dictionary of the History of Technology*,
  (Routledge, 1996)

Milton, Giles, *Big Chief Elizabeth* (Hodder & Stoughton, 2000)

Newman, James R., *The World of Mathematics*
  (George Allen and Unwin Ltd., 1960)

Osborne, Roger, *The Floating Egg* (Pimlico, 1998)

Parker, Derek, *Nell Gwyn* (Sutton Publishing, 2000)

Pepys, S., *Everybody's Pepys* (G. Bell and Sons, 1926)

Petroski, Henry, *The Pencil* (Alfred A. Knopf, 1999)

Porter, Stephen, *The Great Plague* (Sutton Publishing, 1999)

Rackham, Oliver, *Trees & Woodland in the British Landscape*,
  (Phoenix Press, 1976)

Reston, James Jnr., *Galileo* (Cassell Publishers Ltd., 1994)

Ridley, Jasper, *Henry VIII*
  (Fromm International Publishing Corporation, 1986)

Rolt, L.T.C., and Allen, J.S., *The Steam Engine of Thomas Newcomen*
  (Landmark Publishing Ltd., 1997)

Ronan, Colin A., *The Cambridge Illustrated History of World's Science*
  (Book Club Associates, 1983)

Schama, Simon, *A History of Britain – At the Edge of the World
  3000BC – AD1603,* (BBC Worldwide Ltd., 2000)

Schama, Simon, *A History of Britain – The British Wars 1603-1776*
  (BBC Worldwide Ltd., 2001)

Singer, Charles, *A History of Scientific Ideas*,
  (Barnes & Noble Books, 1959)

Singer, Charles, *The Earliest Chemical Industry – An Essay in the Historical
  Relations of Economics & Technology illustrated from the Alum Trade*
  (The Folio Society, 1948)

Summerson, John, *Inigo Jones* (Yale University Press, 2000)

Tinniswood, Adrian, *His Invention so Fertile – a life of Christopher Wren*
  (Jonathan Cape, 2001)

Williamson, David, *The Kings and Queens of England*
  (National Portrait Gallery, 1998)

# Index

Numbers in italics refer to
illustrations.

# Picture Acknowledgements

While every effort has been made to trace copyright holders for illustrations featured in this book, the publishers will be glad to make proper acknowledgements in future editions in the event that any regrettable omissions have occurred at the time of going to press.

Endpapers: The *What the Tudors Did For Us* woodcut was commissioned for the BBC series reproduced courtesy of John Lawrence, also used on pages 12-13. The *What the Stuarts Did For Us* woodcut was also created for the BBC series and is reproduced courtesy of Black and White Line/Dave Hopkins, also used on pages 130-131.

Frontispiece: *The coffee-houses of London were hotbeds of gossip, intellectual discussion and political debate*; reproduced courtesy of Mary Evans Picture Library.

The woodcuts used decoratively throughout the text are from *800 Decorative Woodcuts for Artists and Craftspeople* (Dover Publications, Inc, 1999), originally published as *Specimens of Early Wood Engravings* (William Dodd, 1862).

Diagrams on pages 48, 76,165, 170, 212 and 220 were drawn by Dan Newman, Perfect Bound Ltd.

With thanks to the following for permission to reproduce their images:

**AKG London** 15 (also 237) 20, 25 (also 15 and 236), 30 (also 2), 33 (also 237), 35, 39 (also 2), 45 (also 2), 50, 54, 71 (also 237), 72/3 (also 238), 77, 82, 121, 128, 133 (all except Charles II), 148 (also 239), 171, 178 (also 240), 184, 186, 208, (also 3 and 240)

**Paul Bradshaw** 79

**BBC Publicity** 8, 21, 37, 81,115, 124

**British Library** 86,145

**Corporation of Trinity House** 206

**Historical Portraits Ltd, London, UK/Bridgeman Art Library** 193

**Mary Evans Picture Library** 15 (Edward VI and Henry VII), 23, 53, 88, 97, 99, 108, 115, 116, 119, 120 (also 238), 122, 125, 135, 141 (also 239), 142, 143, 152, 153,189 (also 238), 195 (also 6), 210, 221, 222, 228 (also 241), 229

**Mike Garner** 69

**Museum of London** 139

**National Portrait Gallery Picture Library** 134 (also 133 and 239), 161 (also 3 and 238),180, 203

**National Trust Photo Library/John Hammond** 104

**National Trust Photo Library/Nadia MacKensie** 112

**National Trust Photo Library/Geoff Morgan** 113

**Pete Jones** 41 (also 8)

**Private Collection/Bridgeman Art Library** 190 (also 3 and 240)

**Public Records Office** 84, 150

**Royal Geographical Society, London, UK/Bridgeman Art Library** 95

**NMPFT/ Science and Society Picture Library** 46, 109 (also 238), 158 (also 239), 211, 215 (also 240), 217, 218

**Emma Sutton** 115

**Alom Shaha** 56, 90, 91, 92

**Victoria & Albert Museum Picture Library** 123

All other images reproduced courtesy of Adam Hart-Davis.